Philosophy of Engineering and Technology

Volume 52

The *Philosophy of Engineering and Technology* book series provides the multifaceted and rapidly growing discipline of the philosophy of technology with a central overarching and integrative platform. It publishes on all topics in the philosophy of technology and is open to all research communities across the world.

Specifically, it publishes edited volumes and monographs in:

- the phenomenology, anthropology and socio-politics of technology and engineering
- the emergent fields of the ontology and epistemology of technology and design
- engineering ethics and the ethics of specific technologies ranging from nuclear technologies to artificial intelligence
- written from philosophical and practitioners' perspectives and authored by philosophers and practitioners

The series also welcomes proposals that bring these fields together or advance philosophy of engineering and technology in other integrative ways.

Proposals should include:

- A short synopsis of the work or the introduction chapter
- The proposed Table of Contents
- The CV of the lead author(s) or editor(s)
- If available: one sample chapter

We aim to make a first decision within 1 month of submission. In case of a positive first decision the work will be provisionally contracted; the final decision about publication will depend upon the result of anonymous peer review of the complete manuscript. We aim to have the complete work peer-reviewed within 3 months of submission.

The series discourages the submission of manuscripts that contain reprints of previous published material and/or manuscripts that are below 150 pages/75,000 words.

For inquiries and submission of proposals authors can contact either editor-in-chief: Pieter Vermaas at p.e.vermaas@tudelft.nl, or Darryl Cressman at darryl.cressman@maastrichtuniversity.nl

Levi Checketts
Editor

What Went Wrong with the West: Carl Mitcham's Critique of Western Philosophy of Technology

Springer

Editor
Levi Checketts
Religion and Philosophy; Centre for Applied Ethics
Hong Kong Baptist University
Kowloon, Hong Kong

ISSN 1879-7202 ISSN 1879-7210 (electronic)
Philosophy of Engineering and Technology
ISBN 978-3-032-22213-8 ISBN 978-3-032-22214-5 (eBook)
https://doi.org/10.1007/978-3-032-22214-5

This Springer imprint is published by the registered company Springer Nature Switzerland AG
The registered company address is: Gewerbestrasse 11, 6330 Cham, Switzerland

Introduction

Abstract This introductory chapter lays out a very brief overview of the contributions Carl Mitcham has made to philosophy of technology and especially his engagement with Chinese thought. Rather than an Orientalist or Western chauvinist approach, Mitcham has facilitated dialogue on Chinese philosophy of technology on terms outlined by Chinese scholars themselves. A context of the need for better engagement between Western and Chinese philosophy of technology is followed by discussion of Mitcham's contributions to the field and Mitcham's particular impact on the editor. This introduction concludes with an outline of the chapters in this work.

Keywords Carl Mitcham · Joseph Needham · Philosophy of technology · Chinese philosophy · East-West comparative philosophy

Beyond the West

Philosophy has a generally Western tint to it. Aside from the fact that the name "philosophy" is an artifact from the Athenian tradition, particularly Socrates's own denial of himself as "wise" (i.e., sophist) but rather a lover of wisdom (i.e., philosopher), most genealogies and studies of philosophy assume the source of ancient wisdom is primarily Greco-Roman, continued through the Middle Ages by Christian scholastics, given new breath in the Renaissance and Enlightenment, and now typically divided between (European) "continental" and (Anglophone) "analytic" approaches.

But the Axial Age, the supposed Golden Age of philosophy, was also significant for the emergence of the Upanishads, Daoism, Confucianism, the Hebrew prophetic tradition, and more. In recent years, universities in the West have tried to make up for their occidental bias by offering courses with titles like "world wisdom traditions" or hiring scholars specializing in Chinese or Indian philosophy. Meanwhile, in China, whether that be the once-colonized Hong Kong or the safely monocultural Beijing, philosophers are often working unflinchingly with European traditions, their own historical traditions, or new approaches derived from Maoism. If "philosophy" has historically carried in its name an implicit disdain for non-European "barbarians," the Chinese are not too concerned about this legacy.

Of course, this current flourishing of Chinese thought has its own historical quirk. While China's late nineteenth and early twentieth century history can be read as a continuation of China's problematic relationship with the West, including in the very spirit of 1911 Revolution and the philosophy of Marx guiding the communist party in the Chinese Civil War, the country became closed off to the West once the People's Republic was established. If and when mid to late-twentieth century Western philosophical texts ever discuss China, they do so with a surprising amount of naivete, usually rumors or old accounts from before Mao. It is only starting with Deng Xiaoping's Reform and Opening-Up efforts in the late 1970s that Chinese scholars began to study and work abroad and that Western scholars had any access to the Middle Kingdom. Today, China's universities are globally competitive, and Chinese scholars are on the faculty of every top 50 university, demonstrating clear scientific prowess.

And yet, in spite of this tremendous sequence of events, we read history through a Western hermeneutic lens. We awe at how developing China "caught up" with the West, not how China carved its own path in a tumultuous century of globalization and uniformity. The title of this work is a play on Joseph Needham's famous question, discussed in great detail in numerous chapters here (including specifically Mitcham and Nordmann's), which asked why China fell behind the West in sciences and technologies when it had dominated so much of early history. In response, looking at the past 50 years of the so-called Chinese Miracle, the Chinese may ask why the West is now in such disarray. In 2025, this question is especially pointed, given geo-political, international trade, and artificial intelligence competitions between the USA and China.

In summer 2021, at a time when anti-Asian racism had spiked in the USA due to xenophobic state-sponsored propaganda and the fact that a virus causing a global pandemic had first been discovered in China, the founder of the Society for Philosophy and Technology dared to challenge the narrative of Western philosophy of technology in favor of a Chinese approach during his "Lifetime Achievement" keynote speech. Carl Mitcham has spent much of his adult life in what seems to be a typically Western orientation, including teaching for American and European institutions, forming a "monastery for families" in Kentucky, and developing and promoting different approaches to Western philosophy of technology. However, in recent decades his legacy has shifted across the Pacific and he has spent several semesters teaching in Beijing and lecturing throughout China. He is perhaps in the unique position, having been attentive to the US military-industrial project in its Cold War height and now mentoring young scholars in China, to call for a re-reading of the meaning of philosophy of technology.

This volume serves as a meditation and subversive challenge to the Western tendency in philosophy of technology. It simultaneously serves as a celebration of the thought and legacy of Carl Mitcham, a legacy detailed in the biographies of the authors of this collection. Each person contributing is a friend and collaborator with

Mitcham. Each of us has our own unique history with Carl and reason to contribute to this work. Each of us tries to respond to the call he raises, to think about the history of philosophical attitudes toward technology anew, attentive to our own limitations and cultural biases. This work is meant to be provocative: to provoke conversation such as those that fill these pages and have spanned the pre-writing and post-publication of such a work. To understand the aim of this work, it is first important to understand who Carl Mitcham is, at least to the editor.

Mitcham's Significance for Philosophy of Technology

Carl Mitcham's impact on philosophy generally, and especially specific branches of philosophy, such as philosophy of technology, philosophy of engineering, philosophy's engagement with Science and Technology Studies, and even philosophical-theological studies of technology will outlast even those who have personally learned from him. I may venture to say that he has had a more direct, and likely a more outsized, impact on the field of philosophy of technology than even Martin Heidegger or Jacques Ellul. While he doesn't have a catchy neologism to his name, like von Schomberg's Responsible Research and Innovation, Ihde's postphenomenology, Latour's Actor-Network Theory, or Feenberg's Critical Theory of Technology, his way of approaching the field has shaped most philosophers' understandings of the discipline.

Of course, philosophy is not a competition, and Carl would be the first to point this out. Part of the reason Mitcham's legacy is so extensive is because he has a charism that is supportive of efforts to authentically engage with ideas—even against his own views!—meaning his number of allies is great while his detractors are few. The diversity of views outlined in this collection, to say nothing of the many more who were interested in contributing, is a strong witness to his impact across the globe. Contributing to this work are four Chinese philosophers, some of whom are quite junior in their work, an Indian-American accounting professor who specializes in the thought of Mitcham's mentor Ivan Illich, two German philosophers, an Italian and a Russian philosopher, and me, an American living and working in Hong Kong.

I think Mitcham's legacy will be cemented in the annals of philosophy due to at least three notes about who he is as a scholar and a person. The first, what one would of course expect, is his own philosophical insight. Mitcham's greatest theoretical contribution to the field of philosophy of technology (aside from the notable fact that he *established* the field in founding the Society for Philosophy and Technology and serving as its first president!) is the distinction between humanities philosophy of technology and engineering philosophy of technology (Mitcham, 1994). The two approaches amount to two different ways of understanding technology and its role in the world. While there are not too many philosophers writing on this topic today, it is evident in the most prominent direction of philosophy of technology today, i.e., the field of artificial intelligence ethics, where the approaches of humanities-trained ethicists like Mark Coeckelbergh and Pak-Hang Wong differ significantly from industry-based writers like Timnit Gebru and Margaret Mitchell. Retrospectively, after the current fervor surrounding AI has tempered, one might do an interesting investigation on the very different ways AI ethicists have approached relevant issues on the topic, and perhaps even how different professional organizations like the Association for Computing Machinery and the Society for Philosophy and Technology differ in their discussion of computing ethics.

Second, Carl has been an incredible bibliographer. Most scholars of philosophy of technology begin their career with a copy of Mitcham and Mackey's 1972 *Philosophy and Technology*, a reader that outlines what they believed (and many would accept still) as foundational readings within the field. A decade later, he followed up this text with Jim Grote in *Theology and Technology* (1984), though this latter text has enjoyed less readership. Early on, Carl (1973) saw the importance of keeping an annotated bibliography of different writings in philosophy and technology, and his original *Theology and Technology* contains a 180-page topically organized bibliography. Carl recognizes the importance of diverse approaches, and where someone may want to discuss Latour or Heidegger, Mitcham will add in Ortega Y Gassett, Ellul, Illich, and Leo Strauss. His reading recommendations probably have had a greater impact on young scholars' direction of study than would ever be apparent because of the diverse directions he sends scholars according to their interests.

Finally, as hinted above already, Mitcham has never shied from dirtying his hands in disparate fields. Some of the chapters in this volume allude to this, but to wit, we note that after establishing the Society for Philosophy and Technology, he moved for some years into theological considerations. After returning to philosophy, he engaged with Science, Technology and Society (STS) studies scholars (Mitcham & Cutcliffe, 2001) and was among the first Western philosophers to openly dialogue with Chinese philosophers (see Wang, and Zhang, both this volume). As a result, Mitcham has a palpable presence in history of philosophy of technology, philosophy of engineering, STS, East-West comparative philosophy of technology, and theological studies of technology. These contributions are not negligible, and they all together paint a bigger picture of who he is as a thinker, one not so wed to a cultural or methodological framework to miss out on greater discussions or possibilities.

In Gratitude

My own encounter with Carl is instructive on what kind of a scholar and mentor he is. I discovered an old copy of *Theology and Technology* in the Graduate Theological Union's library as a PhD student and found the essays contained in it fascinating. But having studied some STS, I found myself disagreeing with what Peter-Paul Verbeek would have called the "essentialist" approach of Mitcham and other contributors to the problem of technology. In response, I wrote a paper which I presented at the 2017 SPT meeting in Darmstadt, Germany. I was a very green, very introverted, *theology* scholar arriving at this philosopher's conference knowing only barely Shannon Vallor and Mark Coeckelbergh, and only because of opportunities to meet with them in the San Francisco Bay Area.

After a few sessions at the meeting, I had a chance to meet and talk with Carl at one of the coffee breaks. I told him I had read his book, but I disagreed with some of his ideas, and I would be talking about it in my own talk. I don't know what I expected from him, but I was pleasantly surprised that he thought this should be an interesting contribution. He indeed attended my talk and offered constructive feedback, not attacking my overall premise, but offering some suggestions about further directions for developing my idea. Over the course of the rest of the conference, I had enriching conversations with him, including on a trip to Mainz to visit the Gutenberg Museum, where we discussed whether Gutenberg Bibles could be considered sacred in their own right (Mitcham argued the reproductive aspect of printing meant they could not. I rejected this based on the actual treatment of the artifacts. Eventually, I compromised with his way of thinking and accepted that they could be "sacramental" without being "sacred"). Without question, this friendliness and welcome was instrumental in helping me find a home at SPT.

In the years since, Carl has been incredibly supportive of me as a junior scholar. He helped me connect with various people, including Albert Borgmann and the International Jacques Ellul Society and expand the theological work I have been doing. When I was struggling to find full-time work after finishing my PhD, he gave the sound advice that I should turn my gaze toward Korea and China, leading to my willingness to move to Hong Kong. My first major book publication was a reprinting and updating of his *Theology and Technology*, a project we engaged in after he agreed that the text should be more available (Mitcham et al., 2022). My connection to Carl has also facilitated connections with many more scholars in the field beyond what an introvert like me would ordinarily be comfortable with. I've been pleased to host him in Hong Kong for conferences, where his presence has given gravitas to our smallish university events, and am privileged to be at the helm for this text. The list of what I owe to Carl goes far beyond what I've outlined here and is only cut short out of modesty for Carl (and myself!).

Suffice to say, there are many scholars, young and old, who owe much to Mitcham for their ideas, their research, their networks, and even their status. I hope that when I am in my eighties, I have the same energy and enthusiasm he has to foster new generations of talent, help bridge connections between different groups, and use my position for other people's benefit.

Finally, a last anecdote. In a recent trip to Beijing, I had the privilege of presenting my research to Chinese scholars, including Tom Xiaowei Wang, included in this volume. My paper formally focused on the philosophy of Andrew Feenberg and some reflection from my own recent book. However, in the discussion following the talk, Wang noted that my approach to philosophy of technology drew from many different sources rather than a single dogmatic stream, and so he asked if I am not therefore a Mitchamian in my approach. I had not carefully considered this before, but in reflection, I can think of no better compliment to my research than that my philosophical approach echoes the bricolage, bibliographic style of Carl.

Outline of This Book

This book opens with the work of Carl Mitcham himself, an expanded version of his 2021 Lifetime Achievement keynote speech from the Society for Philosophy and Technology's conference in Lille, France, in 2021. As noted above, Mitcham's significance in philosophy of technology is not to be understated. And yet, his focus now is on the future of philosophy of technology and its increasing importance outside of the Western world. Taking up the famous question of Joseph Needham—why the West developed science (and technology) while China did not—Mitcham reverses the assumption. Pointing out the deeply militaristic history of technological development in the West, Mitcham counters that China has largely focused on technology as it benefits the populace in peace. Thus, Mitcham asks why Western attitudes toward technology should be the prevailing attitudes, especially given current threats to our existence.

The chapters following Mitcham's reflect on his own challenge, contradicting the occidental bias inherent in philosophy of technology. The next two chapters trace out Chinese historical attitudes toward technology. The first of these, by Dong Luo, examines the history of philosophical attitudes toward technology in ancient Chinese thought. He notes that the development of Chinese philosophy toward technology seems to have gone from a rather generous appraisal of technical advance in the Zhou Dynasty (1046–256 BCE) to increasingly harsher critiques through the development of Confucianism, Daoism, and Legalism. Luo's examination of different philosophical considerations of technology reveals a pattern, one which he insightfully ties to the devolution of ancient Chinese political society, from an apparently harmonious dynasty to increasing corruption and war between states.

Tom Xiaowei Wang provides a complementary view to this approach by examining the roots of contemporary Chinese philosophy of technology. Wang highlights at least three major sources for contemporary modern thought. The first of these is an organic Chinese philosophy of technology emerging from mainland Marxist philosophy through the thought of Chen Changshu and the "Dialectics of Nature" Marxist approach, which lends itself more to engagement with Science, Technology and Society approaches to philosophy of technology. Wang also identifies Western analytic philosophy and Western philosophy of technology (e.g., Postphenomenology and Heideggerianism) as two other major directions in contemporary Chinese philosophy of technology. However, as demonstrated by the broader corpus of Wang and others, there is perhaps a fourth direction, indicated by a movement to recover ancient Chinese philosophy as a source for philosophy of technology. These directions suggest the variety and importance of Chinese philosophy of technology in the current historical moment, but Wang personally advocates for a specifically Confucian revival for Chinese philosophy of technology.

Following these historical studies, the next three chapters put Western and Chinese attitudes toward technology in conversation. Opening this section is the work of Pak-Hang Wong, who compares Confucian philosophy of technology with the thought of Ivan Illich, Mitcham's own mentor. Wong highlights parallels between the two schools of thought, as well as areas where they converge. He articulates a plausible approach that synthesizes Illich's "conviviality," which is "autonomous and creative intercourse among persons, and the intercourse of persons with their environment," with the Confucian principles of oneness and non-self-centeredness. Taken in a complementary fashion, Confucian oneness and Illichian conviviality emphasize developing and using technologies in a fashion that respects the interconnected and flourishing human life.

Alfred Nordmann examines the concept of ontological anxiety at root in technical imaginaries in the West and China. Starting with the notion of *Technisierung*, something like "technification," Nordmann asks alongside Gernot Böhme why the West has subjected more and more of the world to a scientific dissection in the name of technology, while China did not. This is especially peculiar given China's early advances in science and technology. Nordmann finds an interesting comparison between the world depicted by Francis Bacon and his Chinese contemporary Song Yingxing. Bacon's world was one that humanity should explore for its own exploitation, while Song's was one that humanity should explore as part of nature. Bacon's own approach to the world came from a type of "innerworldly asceticism," contrasting the internally human with the outer wild. The challenge, then, for overcoming the gap between the external world of subjecting nature to the scientist's bench and a kind of romanticized Orientalism lies in a more Chinese aesthetic approach to technical gadgetry, one that appreciates technology for what it can provide for an enjoyable life, and not one constituted by anxiety.

I also draw on the concept of anxiety to discuss the different technical attitudes between the West and China, but draw from the psychoanalysis of Jacques Lacan. If the West is constituted by a Christian heritage, which authors like Lynn White, Jr. and Jacques Ellul have connected to Western technoprogressivism, then the masochistic anxiety inherent to Christian culture is also part of this quest to dominance. Conversely, anxiety of the most prominent Chinese cultural influences—Buddhism, Confucianism, and Daoism—does not take the same shape, though Confucianism, the most pro-technology ancient Chinese philosophy, gets closest. However, in the current age, this is largely irrelevant as modern China is defined by its anxious relation to the West, a relation that frames contemporary Chinese attitudes to progress distinctively from the past.

In the fourth section of this work, more Western-focused perspectives are raised. Christopher Coenen and Aleksandra Kazakova open this section by asking what is entailed in the cultural imaginary of the "West" and how Marxist thought challenges the notion of Western progressivism. Noting the significance of their own biographical locations, they trace modern constructions of the West and East as Europe and European cultures defined themselves in relation to the enigma of Russia. Before contemporary imaginaries about the "Far East," "the East" meant the power of the Eastern Bloc, that is the Soviet Union and its allies, which inevitably included Mao and the People's Republic of China. Ironically, it is this idea of the "East" that provided impetus for socialist countries to pursue science and technology at an advanced rate through the concept of the "Science-Technological Revolution," a vision of Marxist progressivism, with the Soviet Union early surpassing the USA in the Cold War, and today's China advancing on all fronts both in university and corporate research.

Sajay Samuel confirms my own suggestions about the character of the West by examining the theological roots of Western technology. Samuel argues plausibly that the entire notion of technology is a product of a specifically Roman Catholic theological worldview. Distinct from the Aristotelian metaphysics from which he derives his ideas, Thomas Aquinas divides the efficient cause into an agential and instrumental cause, leading to a specific concept of instrumentality. This owes to the Catholic notion of sacraments, instruments of God's gracious action for humanity. This, argues Samuel, is the paradigm for technology, "the domain of separable and neutral things, conventions, or persons that work on someone or thing at the behest of someone else." Interestingly then, Samuel concludes that the Western pursuit of technology cannot be extracted from Christendom's emphasis on salvation.

Agostino Cera's chapter complements Samuel's by engaging in a reflection on theological anthropologies articulated in various approaches to technology. Like Nordmann's discussion of *Technisierung*, Cera argues our current mode is "dataism," wherein everything has been reduced to data to be analyzed and the flow and exchange of data are paramount. This is a logical extension from earlier technical worldviews, including Cartesianism, Vichianism, and Heideggerism. This advance to this worldview transcends Yuval Harari's notion of "*Homo deus*" and skips rather to "*Techne deus*," technology as our new religious attitude.

This volume concludes with a lengthy interview between Carl Mitcham and Zhang Zhihui. In his own words, Mitcham articulates much of his own philosophical journey to the East, his own account of the history of philosophy of technology, and his own biographical narrative. This is an invaluable resource for philosophers of technology thinking about the past 50 years of the field and new directions for philosophers to consider and is an appropriate bookend to frame the broader significance of Mitcham's work highlighted in these pages.

The field of philosophy of technology is relatively young, with the SPT not yet 50 years old (in 2025). Still, the field has undergone numerous shifts, from an initial condensation of the field in the USA with Mitcham, Feenberg, Ihde, Borgmann, and others, to a shift to the Netherlands with the so-called empirical turn and today a shift to China. The scholars in this volume can be read as part of this shift, but the future of the field is uncertain. Will African philosophy of technology gain prominence in the next 20 years? Should a stronger collaboration with religious positions (e.g., theology) steer philosophical discussion more than it has? Will philosophy of technology become more diffuse like its adjacent field (sometimes seen as a twin, other times as a parent) STS? Regardless of the actual change that will happen, this work reminds us that it should not follow Western imperial myths of progress. As Mitcham cautions in his opening chapter, the future that we share will need to be one where our understanding of our relation to nature and the function of technology are radically different from the violent, exploitative form it takes today.

Department of Religion and Philosophy, Centre for Applied Ethics
Hong Kong Baptist University
Kowloon, Hong Kong

Levi Checketts

References

Checketts, L. (2025). *Poor technology: Artificial intelligence and the experience of poverty*. Fortress Press.

Mitcham, C. (1973, April). Bibliography of the philosophy of technology. *Technology and Culture, 14*(2), Part 2.

Mitcham, C. (1994). *Thinking through technology: The path between engineering and philosophy*. University of Chicago Press.

Mithcam, C., & Cutcliffe, S., (Eds.) (2001). *Visions of STS: Counterpoints in science, technology and society studies*. State University of New York Press.

Mitcham, C., & Grote, J. (1984). *Theology and technology: Essays in Christian analysis and exegesis*. University Press of America.

Mitcham, C., Grote, J., & Checketts, L. (2022) *Theology and technology*, 2 Volumes. Wipf&Stock.

Mitcham, C., & Mackey, R. (1972). *Philosophy and technology: Readings in the philosophical problems of technology*. The Free Press.

Contents

Part I
Mitcham's SPT Lifetime Achievement Keynote

Chapter 1
Re-Imagining Chinese Technological Imaginaries

Carl Mitcham

Abstract This article was originally presented as a plenary talk at the Society for Philosophy and Technology 2021 biannual international conference hosted at Lille, France, on the theme "Technological Imaginaries" and retains some of its informal character. The argument is simply that Chinese imaginaries of artifice, emerging independently of the Euro-American—and with a deeper, more continuous history—have a power to invite particularly English-speaking poets and philosophers to reconsider the frames in which art, engineering, technology, politics, theology, and philosophy have been customarily understood.

Keywords Joseph Needham · Technological imaginaries · Philosophy of engineering · Comparative philosophy · Political philosophy

1.1 Introduction

In the spring of 2012, while teaching a course that would take Colorado School of Mines students to China during the summer for a 2-month study abroad, I was visiting the Freer Gallery in the National Museum of Asian Art on the Washington, DC, Mall. After some hours with a collection of ancient Chinese jade and bronze artifacts from the late Neolithic to the Han Dynasty, in the bookshop I stumbled on Deborah Fallows' recently published *Dreaming in Chinese* (2010). Her account of learning the language and culture during an early 2000s residency in Shanghai, with her *Atlantic* correspondent husband James, turned out to be a useful addition to my syllabus—but also marginalized material culture. The chapter on "The Essence of

C. Mitcham (✉)
Colorado School of Mines, Golden, CO, USA
e-mail: cmitcham@mines.edu

L. Checketts (ed.), *What Went Wrong with the West: Carl Mitcham's Critique of Western Philosophy of Technology*, Philosophy of Engineering and Technology 52, https://doi.org/10.1007/978-3-032-22214-5_1

Being Chinese" was focused on the essence of Chinese *hanzi* or characters, without referencing the tools of paper, brush, and ink integral to the art of calligraphy.

Without ever mentioning him, Fallows echoed the imagination of Ezra Pound, the great twentieth century American poet who discovered in Chinese literature a font of imagist poetry—and became a life-long, if eccentric, interpreter of the Chinese canon. Pound's first collection of Chinese poetry, *Cathay*, published in 1915, including a few selections from the *Shijing* 诗经 as well as the great poets Li Bai and Du Fu, inspired T.S. Eliot to declare him the "inventor of Chinese poetry for our time" (Eliot, 1928, p. 14).

That "our time" acknowledged that China has been repeatedly invented by the Western imagination. This re-imagining has not always included adequate respect for the imaginative lives of Chinese themselves. Alas, I will wholly not escape this inadequacy. Orientalism is a constant temptation.

Nevertheless, for some years now I have been in fortunate dialogue with Chinese colleagues across our respective cultures or imaginaries—to adopt an anthropological term now displacing "culture" or "cultural beliefs"—seeking to reconsider one particular inadequacy: that of the Western, especially American, ignorance of Chinese history and philosophy. Given my beliefs regarding the centrality of engineering and technology in modern western history and culture, the exchange has taken a distinctive cast. I have often found myself interrogating a popular trope that is dismissive of Chinese science, engineering, and technology.

Today, of course, this trope has taken on newly aggressive expressions, born of competition if not resentment and fear. The Chinese are being unfair in adopting policies long pursued by the West. As Peter Sloterdijk has quipped, we in the West were all in favor of globalization while we were doing it to others, but not so much when others start doing it to us. I will, however, ignore present day political imaginaries in preference for a longer term perspective.

1.2 In the Shadow of Joseph Needham

In this perspective, we live in debt to Cambridge biochemist become sinologist Joseph Needham. Beginning in the 1930s, Needham was introduced to the Chinese language, science, and imaginary by his student, mistress, collaborator, and eventually second wife, Lu Gwei-djen. With Lu Gwei-djen's tutelage Needham became one of the great self-educated sinologues. His massive multi-volume *Science and Civilization in China* was initially constructed around the question of why China, which for the preceding 2000 years had remained the most technically advanced in the world, around 1500 began to fall behind Europe in technoscientific development. It is a question that has become known eponymously as "the Needham problem." It is, I might add, a contested problem.

As Needham put it, until the latter half of the Ming Dynasty, China had exhibited a level of continuous technoscientific creativity unmatched in human history: roughly 100 significant discoveries and inventions per century for 20 centuries. We all know about papermaking, printing, gunpowder, and the compass. But the list includes a host of other discoveries and inventions: acupuncture, sericulture, tea, umbrella, alcohol, row cropping, seed drill, wheelbarrow, kite, seismometer, rockets, porcelain, mechanical clock, paper money, and the toothbrush—for a casual list scattered over more than two millennia.

Needham nevertheless imagined as a given that European science—in which he included engineering and technology—as it emerged post-Renaissance, was an essential advance in human history, destined to be adopted by the rest of the world. As the first head of the Natural Sciences Section at UNESCO, Needham was the one who inserted "Scientific" to what had originally been simply the "United Nations Educational and Cultural Organization." For Needham science *was* culture—and at the core of any true education. And while honoring and calling attention to its great premodern heritage, he saw China as now deficient—deficient even in appreciating its own past technical contributions to world history.

A story has it that in 1964 when Premier Zhou Enlai met Needham and learned of his project, Zhou was disconcerted about how little Chinese scholars were contributing. In response, he directed the Chinese Academy of Science, which did not then have an Institute for the History of Natural Science, to begin to rectify the situation.

Needham's answer to his own question about the factors contributing to the distinctive achievements and failures of technoscientific development in China appealed to multivariate influences:

- a Confucian social ontology that privileged harmony,
- a Daoist interest in nature that promoted extensive empirical research but practiced *wuwei* 無為 (effortless action),
- the bureaucratic feudalism of Chinese politics, and
- non-capitalist economic formations.

He nevertheless—and I quote—"confidently [believed] that the analyzable differences in social and economic patterns between China and Western Europe will in the end illuminate, as far as anything can ever throw light on it, both the earlier predominance of Chinese science and technology and, also, the later rise of modern science in Europe alone" (Needham, 1969, p. 217).

The study of science and civilization in China was for Needham an exercise in self-knowledge.

1.3 Stretching Needham

Science and Civilization in China—at more than 27 large volumes, with more still in progress—has become the anchor for an expanded project of global scholarly collaboration, now flowing out of and in to the Needham Research Institute at Cambridge University. Although not commonly described as such, Needham also deserves to be recognized as a philosopher of engineering and technology, West and East—a stretch I'm obviously seeking to make.

Modern China tends to be presented, even by Needham, as deficient with regard to many things found in the West. For Needham, the West ultimately remained a norm from which to reach out to China. For a provisional moment here in this presentation, I want to destabilize Needham with a kind of *Gedankenexperiment* that takes China as a norm from which to reflect back on the West.

This thought experiment in normative reversal is embedded in my persistent philosophical questioning of the western imaginary, especially its dominating commitment to engineering and technology. I want to imagine looking at the West and its technological achievements from the perspective of a western academic, an American academic, who has tried to let himself be tutored late in life by China. How might a Chinese philosophy of engineering and technology help me better understand and criticize the American way with technology?

I hasten to add both that I am not anything approaching a well-educated student, and that the hypothetical perspective here is not one that Chinese have adopted. Indeed, it is not one that a Chinese scholar even could adopt. It is an idiosyncratic effort by an amateur—one sometimes described as a dilettante—to imagine Chinese technological imaginaries in ways that might fruitfully re-inform a Euro-American technological imaginary.

1.4 What the West Lacks

In this experiment in normative reversal, let me then selectively tick off four overlapping features of the Chinese technological imaginary that are missing in the West.

First, the European philosophical tradition is deficient in recognizing the centrality of the family and of agriculture. In the West, the model for the state is not the family, as it is in China, but military command or commercial trade.

The *Lushi Chunqiu* 吕氏春秋 (or *Spring and Autumn Annals of Scholar Lu*) includes a section describing the imaginary of those engaged in the "root" occupation of farming with that of merchants. Farmers live on the land, which cannot be moved; they do not easily abandon it in the face of danger. Merchants or traders flee when dangers appear. Farmers are more honest and loyal than merchants and thus exemplify a superior way of life. Such a straightforward appreciation of peasant life is lacking in classical western philosophy. It takes only distorted form in aristocratic theories of household management and modern idealizations of the noble savage.

As the Chinese historian of philosophy Feng Youlan observed,

> The Greeks lived in a maritime country and maintained their prosperity through commerce. They were primarily merchants. And what merchants have to deal with first are the abstract numbers used in their commercial accounts, and only then with concrete things that may be immediately apprehended through these numbers.… [Merchants also] live together in towns [and their social order is based not] on the common interest of the family so much as on [contractual relations]. (Feng, 1948, 25)

The West mistakenly privileges rational contracts among rationalist individuals over embedded, preliterate, subsistence relationships.

Second, reinforcing this abnormal, the founding texts of the West lack a sense of and respect for the quotidian. Books such as the *Iliad*, *Odyssey*, and Hebrew Scriptures are composed around a fantastic male warrior and heroic imaginary devoid of daily life. The pleasures and sorrows of domesticity, both feminine and masculine, are almost wholly missing.

By contrast, the anchoring text of the Chinese imaginary is the *Shijing* or *Book of Odes*, a collection of 305 poems that became stabilized earlier than either the Greek epics or Hebrew Scriptures. Over and over in what has become persistent imagery, it describes and celebrates embodied pleasures and perennial sorrows of common home making, love, and loss.

Take the second ode, "Ge tan" (*Odes of Zhou and the South*). Its first stanza describes a creeping, kudzu-like vine that covers the valley, the leaves of which can also be used to make cloth, and the enhancing presence of birds alive in the bushes and trees. In the second stanza, the poet cuts and boils the leaves "to make cloth fine and coarse, for clothes that will please me." In a final stanza the female narrator describes washing her dress and returning home to comfort her parents: technics and domesticity entwined.

According to the contemporary philosopher Li Zehou, Western philosophy lacks such a domesticated one-world ontology. Instead, the Western imaginary is bifurcated into this world and a transcendent, so that human beings find themselves at a loss, traveling like St. Augustine to Carthage, burning in a cauldron of illicit loves, abandoning his unnamed mistress while searching for some supernatural home. The Chinese imaginary, by contrast, is grounded in a this-world ontology and the prominence of pleasures rooted in utilitarian materiality.

Li (1988) sees the distinctive Chinese aesthetic imaginary disclosed in the graph of the very word for beauty, 美 *mei*, which is a compound of two characters 羊 *yang* (sheep) above and 大 *da* (big) below: thus imaging "big sheep"—the type of animal that would make any farming family happy and serve as well as a proper sacrifice in a shamanist ritual.

The Chinese domestic imaginary is at home in the world.

Third, the mythologies of the West are dominated by violence. The Hebrew god has a chosen people, slaughters the non-chosen to protect them, and severely punishes the chosen when they disobey his dictates. The Greek gods lead dissolute, warring lives, and Prometheus has to steal fire to protect humans from the stupidity of his brother Epimetheus.

By contrast, in Chinese mythology, as Yuk Hui has pointed out in *The Question Concerning Technology in China*,

> we find another mythology concerning the creation of human beings and the origin of technics, one in which there is no Promethean figure. It tells instead of three ancient emperors ...: Fuxi (伏羲), Nüwa (女娲) and Shennong (神農). The female goddess Nüwa, who is represented as a half-human, half-snake figure, created human beings from clay. Nüwa's brother, and later husband, is Fuxi, a half-dragon, half-human figure who invented [the tri-

> grams of the *Yijing*].... Shennong, which literally means "divine farmer," is also the god of fire, and after his death becomes the god of the kitchen....

There is no comparable world affirming mythology in the West. As Yuk Hui concluded,

> Unlike Greek mythology, then, in which the Titan revolted against the gods by bestowing fire and means of subsistence upon human beings, thus raising them above animals, in Chinese mythology there was no such rebellion and no such transcendence ...; this endowment is seen instead as owing to the benevolence of the ancient sages. (Hui, 2016, 14–16)

In the West the gods are alien and oppressive, present within the household only when imported or imposed from without. There are no kitchen gods to help cultivate the technics of the table.

Fourth, the Western imaginary lacks substantive historico-cultural continuity. This is most obviously the case in the history of Western philosophy, which is riven with breaks and oppositions such as Greek versus Christian, ancient-medieval-modern, analytic versus continental, ethics versus metaphysics versus epistemology. The history of Chinese philosophy, by contrast, is Daoist *and* Confucian *and* Buddhist all the way down—with one written language all the way back.

Moreover, in Daoism, Chinese philosophy maintains links to a shamanist autochthonic. This is an attachment conspicuous by its absence in the West, after Greek philosophers severed *logos* from *mythos* and Christians followed by divinizing *logos*. This loss of the indigenous preliterate is even more pronounced in the America techno-economic libertarian imaginary than in Europe.

1.5 An Imaginary of Engineerings

These four Western deficiencies—lack of agrarian life, split world ontology, violent gods, and historical discontinuity—find expression in a contrast of engineering imaginaries, East and West. The Chinese term commonly translated as "engineering" is *gongcheng* 工程. In Needham's etymology, the first character *gong* 工 represents a carpenter's square and "from the earliest times ... implied work of an artisanal character"; the second character *cheng* 程 "originally meant measurement, dimension, quantity, rule, examination, reckoning, etc." (Needham & Wang, 1965, p. 9). In the West, as Needham notes, engineering emerged as a distinctive way of building in the military; "not till the 18th century was it freed from its primary military connotation" (ibid.). But even when freed, we might add, Western engineering has remained infused with a conquering spirit.

By contrast, "the associations of the Chinese terms for engineering and artisans seem always to have been more civilian and less military than those of the [war making] '*ingeniarii*' ... of the West" (*Ibid*, p. 10). The legendary founding king of the third millennium BCE Xia Dynasty was mythologized as *Dà Yŭ zhìshuĭ* 大禹治水 "Great Water Ruling Yu" for his high moral character and introduction of flood management along the Huang He or Yellow River. The other two great proto-engineering projects of Chinese antiquity were likewise civilian oriented (the Grand Canal) or for military defense rather than aggression (the Great Wall).

Even when China began to adopt the engineering techniques of the West during the Self-Strengthening Movement of the late Qing Dynasty, the guiding motto was *Zhōngxué wèi tǐ, xīxué wèi yòng* 中学为体，西学为用 "Chinese learning for substance, Western learning for application." Engineering was, we might say, repurposed as fundamentally defensive rather than conquest oriented: a means to protect against imperialist barbarians demanding free trade commercial rights to market opium in China. In our colleague Li Bocong's (2021) contemporary philosophy of engineering, too, engineering remains primarily defensive and imagined as ultimately cultivating harmony between heaven and earth.

The tragedy is that China does not always seem able to keep engineering for defensive application and free from the cultural corruptions of an inner alliance with capitalism. Or is this suspicion only present in the distorting mirror of my Orientalism?

Consider, finally, a suggestive comparison of imaginary mythic elements present in American and Chinese space programs, as analyzed by our colleagues Glen Miller, Michael Portal, and XU Xin (2023). In 1960 as American space policy entrepreneurs began to imagine a project to counter Soviet achievements and land a man on the Moon, NASA manager Abe Silverstein gave the mission its cosmo-technical name because he thought that "Apollo riding his chariot across the sun was appropriate to the grand scale of the proposed program." Yet any continuity between the worlds of the Greek myth and American life is vanishingly thin; Apollo was simply wheeled in from outside, as it were, and pasted on a rocket engineering with roots in Nazi Germany—a *machina ex deo*, if ever there was one.

By contrast, the contemporary Chinese lunar exploration space craft are named after *Cháng'é* 嫦娥, a mythic goddess of the Moon, who is remembered each year even today during the Mid-Autumn Festival, with the sharing of special Moon cake pastries. Not even in Greece is there still any annual festival of Apollo.

As Miller and co-authors further observe, in America the mythic elements are really the engineers themselves. "Many American myths originate in actual engineering and technological accomplishments that seem miraculous to most people and are characterized as decisive breaks from the world as it had been known." Rather than the sage king Great Yu, Americans genuflect before the innovating billionaires Steve Jobs, Bill Gates, Jeff Bezos, and Elon Musk.

1.6 From Imaginary to Political Power

From its earliest days, western Sinology has oscillated between finding in China answers to its own questions (Leibniz and Voltaire) and seeing it as inherently deficient (Matteo Ricci and Hegel). Needham's sinology synthesized the two: Sinology became Sinophilia. No one before had ever undertaken such broad studies of the deep Chinese past while enthusiastically engaging with the transforming Chinese present. Needham arguably embodied the best features of a European colonialism in a counter-Orientalist imaginary.

More richly than any of his sinological predecessors, Needham saw both European and Chinese civilizations as constellations of science, technology, and society. Needham was doing Science, Technology, and Society (STS) studies at least half a century before more conventional academics invented it. One might even propose that Needham's *Science and Civilization in China* turned sinology into STS—but as Science, Technology, and Society not Science and Technology Studies.

Then along came the Americans. During the rise of what Henry Luce would imagine as "the American Century"—and particularly in conjunction with plans during the Second World War to incorporate China into an international anti-Communist hegemony—the United States military and foreign affairs establishments sponsored creation of a replacement for Sinology: China Studies. This military-commercial imaginary aimed to construct China as the post-war American political power anchor in Asia.

China, however, rejected this role in favor of its own trajectory of experiments in techno-political power imaginaries—a trajectory that can be broadly parsed into the regimes of Mao Zedong, Deng Xiaoping, and now Xi Jinping. China Studies likewise wriggled out from under its initial military-political patronage. Many of those whom the U.S. government posted in Hong Kong and Taiwan to learn Chinese and become "China watchers" wound up becoming attracted to Chinese history, culture, and philosophy for their own sakes.

Confucianism especially slowly garnered respect in the West and has occasionally been promoted (e.g., by Roger Ames at the journal *Philosophy East and West*, 1951-present) as an alternative to the Western philosophical tradition and complement to the ideology of Chinese Marxism. Chinese studies as taught in American universities is now something of a mangle of classical Chinese literature, Needham's sinology, international relations, cultural studies, and philosophy—along with a small dose of STS. Across various post-war American imaginations, China has often functioned for those who would look at it carefully, as a mirror for the West.

1.7 China as Mirror

The suggestion of China as mirror recalls the *zhao yaojing* 照妖镜 or magic monster-revealing speculum that accompanied the monk Xuanzang on his legendary seventh century *Xīyóu jì* 西遊記 *Journey to the West* to collect sacred Buddhist sutras. Let me now venture a concluding, speculative gaze into a philosophical version of a Chinese mirror that might accompany a *Journey to the East*.

At least since Rousseau, one persistent if subaltern theme in European philosophy has been a criticism of modernity that calls for some kind of return to the ancients. From Nietzsche and Heidegger this trope has become increasingly prominent, with Leo Strauss giving it a distinctive political philosophical articulation.

Roger Ames, as mentioned, gives it a Chinese twist by offering Confucian role morality as an alternative to what he, along with Alasdair MacIntyre and others, have argued to be fundamental weaknesses in Euro-American ethics discourse.

For Strauss, however, the fundamental issue is neither ontology nor ethics but politics, manifested in what he calls the theological-political problem. In Strauss this problem has multiple dimensions, but one is that Enlightenment philosophy, as in Spinoza's *Theological-Political Treatise*, prosecutes a radical criticism of religious belief—a criticism that at once opens doors to the modern technological project *and* destabilizes any political order the project would ostensibly benefit.

According to Strauss (1967), the dynamism of Western civilization is grounded in a dialectic between two imaginaries: the Bible and Greek philosophy, revelation and reason. Strauss considers how in the political orders in which all three Abrahamic religions (Judaism, Christianity, and Islam) have been manifest, this dialectic introduced into political affairs a distorting fanaticism with which philosophy has been forced to contend. In Western European modernity this led to promoting the scientific and technological conquest of nature as an alternative to persistent internecine religious warfare. The political theology of the modern project nevertheless seems to lead to a theologizing of politics or technoscience or both—and there through to potential technological self-destruction or founding of the universal and homogeneous regime: globalization on steroids.

The religious beliefs that Spinoza attacks are, however, those peculiar to the Abrahamic religions and their claims to monotheistic revelation. These monotheistic claims, as the Egyptologist Jan Assmann (2007) has argued, constitute an abandonment of cosmotheism for an imaginary of what he calls "secondary religion". The rebellion against cosmotheism and claim to absolute truth introduces into human affairs an extremism or intolerance of the natural contentions within the *polis* over different visions of the good life. What is revealed by God is not subject to moderation or the quotidian give-and-take negotiations of political discourse. Even when the religious claims to revelation are challenged, the challenging itself is often undertaken with a kind of zealotry that echoes the one it questions: one imaginary reflecting another. We can see this reproduction in a persistent Enlightenment presumption for technocracy. The solution to all problems is no longer an appeal to an all-powerful God but to technoscientific innovation.

While arguing that modern political thought in the West has reached a cul-de-sac that requires reconsideration of the ancients, Strauss never made any serious effort to consider Chinese or any premodern non-Western tradition of political philosophy. What is unspoken in Strauss is that there exists even today an alternative political imaginary absent the distorting infusion of monotheistic revelation, one that can be detected in China (via some hermeneutic sympathies) with its own quietly different dialectic of Daoism, Confucianism, and Buddhism and a possibly distinctive trajectory of scientific and technological development valorizing more harmony than conquest. There has not been in China, for instance, any fanatic opposition to heliocentric astronomy or to Darwinian evolution, much less to covid-19 vaccination—nor the precise zealotry for science and engineering as often manifested in the American imaginary.

What Needham conceptualized in problematic historical terms (Why did China after 2000 years of technological superiority to the West not develop modern science?), the political philosophy of Strauss invites re-interpreting differently as a fanaticizing secularization of monotheism. One might also interpret the Maoist fanaticism that prioritized cadre loyalty and discipline over expertise—scientific, medical, engineering, or technological—in similar terms. Yet as is the case with Western religious anti-technoscientic practices, such zealotry tends toward the self-delimiting. The masses don't really want the masses to build dams by hand—although they can be seduced for a time to humor ideologues who do. The abundant scar literature provides sobering testimony to another Assmann concept, that of cultural memory. In speaking of a Chinese technological imaginary, we might cautiously give some regimes priority over others.

1.8 Conclusion

This has been a big-picture, broad-brush talk on an insufficiently refined effort to think our being-with technology today: Not the kind customary in philosophy and technology studies. It does not lend itself to any deft summary. I have furthermore ventured generalizations and imaginings that colleagues both West and East may find presumptuous, weak, or objectionable. I trust that, East and West, you will forgive and correct my breach of scholarly sobriety. In this spirit I welcome questions, comments, and criticisms.

References

Assmann, J. (2007). *The price of monotheism*. Stanford University Press.

Eliot, T. S. (Ed.). (1928). *Ezra Pound: Selected poems*. Feber & Faber.

Fallows, D. (2010). *Dreaming in Chinese: Mandarin lessons in life, love, and language*. Bloomsbury.

Feng, Y. (1948). *A short history of Chinese philosophy*. Collier-Macmillan.

Hui, Y. (2016). *The question concerning technology in China: An essay in cosmotechnics*. Urbanomic.

Li, Z. (1988). *The path of beauty: A study of Chinese aesthetics*. Oxford University Press.

Li, B. (2021). *An introduction to the philosophy of engineering: I create, therefore I am* (S. Zhang & N. Wang, Trans.). Springer Nature.

Miller, G., Portal, M., & Xu, X. (2023). Engineering myth in China and the United States. In S. H. Christensen, A. Buch, E. Conlon, C. Didier, C. Mitcham, & M. Murphy (Eds.), *Engineering, social sciences, and the humanities: Have their conversations come of age?* (pp. 341–360). Springer Nature. https://doi.org/10.1007/978-3-031-11601-8_17

Needham, J. (1969). *The grand titration: Science and society in East and West*. Allen & Unwin.

Needham, J., & Wang, L. (1965). *Science and civilization in China, Vol. 4, part 2: Physics and physical technology, mechanical engineering*. Cambridge University Press.
Pound, E. (1915). *Cathay*. Elkin Mathews.
Strauss, L. (1967). *Jerusalem and Athens. Some preliminary reflections*. Library, City University of New York.

Part II
Chinese Philosophies of Technology

Chapter 2
Shifts of Technological Understanding in Pre-Qin Chinese Thought: Social-Political Relevance and Implications

Dong Luo

Abstract Inquiries into traditional Chinese technological thought typically concentrate on the philosophical insights within Confucianism and Taoism. I explore the overlooked aspects of technological thought within Legalism. And by engaging in a comparative analysis with the technological ideas expressed in *The Rites of Zhou*, the teachings of Confucius, the ideas of Lord Shang, and the philosophical insights of *Zhuangzi*, we can discern the evolution of technological understanding in pre-Qin Chinese thought. These shifts in technological perspective are not only reflective of the profound social-political changes of the time but also carry significant implications for the socio-political dimensions of traditional Chinese culture.

Keywords Technological understanding · Legalism · *The Rites of Zhou* · Social-political relevance and implications

2.1 Introduction

On one hand, philosophers of technology critique dominant Western technological paradigms, prompting researchers to seek insights from Eastern perspectives, including ancient Chinese perspectives. On the other hand, for many observers, China's rapid modernization in the last 40 years, underscores the need to comprehend Chinese technological understanding for social, political, and historical reasons. The potential presence of distinct characteristics within traditional Chinese

D. Luo (✉)
Institute for Advanced Studies in Philosophy, Science, and Technology, South China University of Technology, Guangzhou, China
e-mail: luodong@scut.edu.cn; luodong12b@mails.ucas.ac.cn

L. Checketts (ed.), *What Went Wrong with the West: Carl Mitcham's Critique of Western Philosophy of Technology*, Philosophy of Engineering and Technology 52, https://doi.org/10.1007/978-3-032-22214-5_2

thought that might have contributed to China's rapid modernization sparks growing interest in ancient Chinese technological thought.

Studies of technological thought in Chinese philosophy typically follow two main approaches.The first one utilizes Western concepts like the *Technê* and *aretê* from ancient Greece, often drawing further inspiration from modern thinkers like Heidegger on the nature of Being. The other approach focuses on the *Dao-Qi* (道-器) relation within Confucian or Daoist frameworks.

In addition to Confucianism and Daoism, the school of Legalism has also profoundly influenced traditional Chinese thought. Some major Confucian philosophers were perceived as amalgamating Confucianism and Legalism, presenting Confucianism superficially while embodying Legalism at its core (Tai, 2008). However, the technological thought of Legalism and its possible influence on Chinese understanding of technology have not been thoroughly explored yet. Comparing the skeptical attitudes toward techniques by Confucianism and Daoism, Legalism, especially the Legalism classic, *The Book of Lord Shang*, espouses a very strong hostility toward craftsmen. These differing attitudes toward techniques among Confucians, Taoists, and Legalists suggest nuanced technological thoughts within ancient Chinese philosophies.

Moreover, even predating Confucianism and Daoism, early Chinese texts like *The Rites of Zhou* articulate a clear pro-technology stance, which takes techniques to be inventions of Sages with an understanding of *Dao*. This promotes a "sacred" view of technology in the early period of Zhou Dynasty (1046–256 B.C.). From this pro-technology stance in early texts to the growing unfriendly or hostile attitudes towards technology from Confucianism to Daoism, and to Legalism, shifts in ancient Chinese technological thought over time are indicated.

However, these possible shifts have been largely neglected. I shall examine often neglected aspects of technological thought in Legalism and examine the relevance of the shifts of technological thought in pre-Qin Chinese thought with the changes in the social-political contexts over that period, and how the changes in the social-political contexts might affected the shift of technological thought in ancient China before Qin Dynasty (221–206 B.C.), and discuss the implication of this relevance.

2.2 Technological Thought from a Confucian Perspective

The *Dao-Qi* relation stands as a focal point in the discourse surrounding ancient Chinese technological thought, as extensively explored in existing literature (Hui, 2016). In ancient Chinese philosophy, the concept of techniques is perceived as the unification of *Qi* (器, which can be interpreted as "tools," "utensils," or "technical objects") and *Dao* (道). According to the Confucianism classic *The Book of Changes*,

> May we not say that *Qian* and *Kun* [= *Yin* and *Yang*] are the secret and essence of the *Change*? *Qian* and *Kun* being established in their places, the system of *Changes* was thereby constituted...that which is antecedent to the material form exists, we say, as *Dao*,

> and that which is subsequent to the material form exists, we say, as *Qi*...The successive movement of *Yin* and *Yang* constitutes what is called the *Dao*...if there is an appearance, then we call it *phenomena* (象); if there is a [material] form, we call it *Qi*. (*The Book of Changes*, 2011)[1]

Within this worldview, the universe is characterized by the perpetual movement and transformation of *Yin* and *Yang* which which can be roughly understood as the active and passive aspects of nature respectively. The paths and patterns of these movements and transformations are termed *Dao*. What individuals perceive within these movements are labeled as *phenomena*. The material objects, including technical objects, which are born from *Dao* or successive movements of *Yin* and *Yang* and manifest specific *phenomena* or forms to people under specific circumstances, are called *Qi*. *Qi* assume distinct forms, representing the *phenomena* of *Dao under specific conditions*. Artifacts or utensils, as *instances of Qi*, are crafted by individuals to *serve particular purposes* (Hui, 2016, p. 100).

In *The Analects of Confucius*, Confucius (551–479 B.C.) articulates that "A gentleman should not become a utensil."(君子不器) (Confucius, 2006).[2] Through a dialogue with his pupil Zi Gong, Confucius elaborates on this analogy. "Zi Gong asked, 'What do you say of me?' The Master said, 'You are a utensil.' 'What is a utensil?' 'A gemmed sacrificial utensil,'" (Confucius, 2006). A utensil represents a specific purpose here, implying limitations and inflexibility. It contrasts with the ideal of a gentleman in Confucian philosophy. Confucian ideals emphasize a gentleman's adaptability to changing situations, contrasting sharply with the fixed purpose within a utensil (Zhu Xi, 2012).

Interpreting the *Dao-Qi* relation through this lens, Confucius' words suggest that while a utensil is bound to a singular or specific purpose, *Dao*, with its boundless potential, can assume myriad forms under diverse circumstances. Hence, a gentleman who comprehends and conforms to *Dao* should exhibit flexibility in both thought and action, adept at navigating varied situations. *Dao*, devoid of fixed form, can engender diverse *Qi*. A gentleman, unbound by any singular purpose, possesses unrestricted adaptability to various purposes or causes, thereby embodying the notion that anything is possible.

It's worth noting that Confucius' stance does not imply a wholly negative view towards *Qi* or technological objects. In another passage, he advises, "The mechanic or craftsman, who wishes to do his or her work well, must first sharpen his or her tools," (Confucius, 2006). Hence, *Qi*, bounded to specific purposes, remains essential for fulfilling those purposes effectively.

In short, while acknowledging the importance of *Qi* in achieving specific goals, Confucius places greater emphasis on formless *Dao*, suggesting its elevated significance. In essence, an accomplished scholar or a gentleman is encouraged not

---

[1] The English renderings of the Chinese classic texts presented in this essay are by myself. My translation of this particular paragraph based on the translation of James Legge (2009).

[2] I appreciate Andrea Breard for discussions on how to translate this line.

to confine himself or herself to the limitations like a utensil but to embody the boundless potential of *Dao*.

This perspective does not diminish individuals who align themselves with specific roles; rather, it serves as an exhortation for personal growth and intellectual attainment. For instance, when Confucius remarks "You are a utensil" to Zi Gong, he does not disparage him but urges him towards becoming a gentleman.

2.3 Technological Thought from the Taoism Perspective

Discussions on Taoist technological thought also often revolve around the *Dao-Qi* relationship, exemplified by the story of butcher Pao Ding in *Zhuangzi*, a work by the philosopher Zhuangzi (369–286 B.C.), who lived over 150 years after Confucius. When questioned by Prince Wen Hui about his exceptional butchering skills, Pao Ding responded:

> What your servant is good at is in advance of any art. When I first began to cut up an ox, I saw nothing but the whole carcase. After three years I no longer see it as a whole, but in parts. Now I work on it by intuition, and do not look at it with my eyes. The use of my senses is discarded, and my spirit acts as it wills. Following the natural structure of oxen, my knife slips through the great crevices and slides through the great cavities, taking advantage of the facilities thus presented. My art avoids the membranous ligatures, and much more the great bones. A good cook replaces his knife every year[-it may have been injured in cutting]; an ordinary cook replaces his knife every month [-it may have been broken]. Now my knife has been in use for nineteen years. It has cut up thousands of oxen, and yet its edge is as sharp as if it had newly come from the whetstone… (Zhuangzi, "Nourishing the Lord of Life", 2006)

This narrative is often interpreted as emphasizing the primacy of *Dao* over artifacts or skills in achieving perfection. It illustrates the unification of *Qi* and *Dao* in technique: possessing a good *Qi* (tool), such as a good butchering knife, is insufficient; understanding *Dao*, such as the *way* of butchering, is essential and holds greater significance for mastery in a craft (Hui, 2016, pp.103–104).

While both Confucianism and Taoism advocate the unification of *Qi* and *Dao*, their approaches diverge significantly. As discussed in Sect. 2.2, Confucius views the transition from *Dao* to *Qi* as a loss of freedom. *Qi* represents a specific manifestation of the formless *Dao*, bounded by particular circumstances. However, achieving a purpose effectively necessitates possessing a good tool. In contrast, Zhuangzi contends that, even for tasks with singular purposes, the environments or circumstances for fulfilling those purposes may vary. Consequently, it is crucial for a craftsman to comprehend the *formless Dao* even for the singular purpose and be capable of adapting to *dynamic* circumstances within the *Dao* for the specific purpose. Hence, for Zhuangzi, unlike Confucius, the focus lies not in finding the best tool but in discovering the best *way*, or *Dao*, to achieve *the specific purpose*.

Confucian discourse on the *Dao-Qi* relation hints a *predetermined* approach to fulfilling a purpose. He emphasizes finding the right tool to achieve the purpose, adhering to established principles. In contrast, Zhuangzi's focus lies in discovering the optimal *Dao*, or the most effective way, to achieve a purpose, prioritizing this over the selection of a specific tool tied to *a particular Dao* or way.

To illustrate this, consider the story of the butcher Pao Ding. Pao Ding, a good cook, and an ordinary cook employ different methods of butchering. Unlike the other two types of cooks, Pao Ding's ox butchering does not simply rely on his tools. Instead, he prioritized understanding the *Dao* of butchering—the underlying principles of butchering or the best way of butchering. As he described, "Following the *natural* structure of oxen, my knife slips through the great crevices and slides through the great cavities… avoids the membranous ligatures, and much more the great bones." This mastery of the *Dao*, or way, allowed Pao Ding to excel with the same tool under various circumstances, as evidenced by, throughout his career, his butchering knife "has cut up thousands of oxen, and yet its edge is as sharp as if it had newly come from the whetstone."

In contrast, an ordinary cook, lacking the understanding of the *Dao* of butchering, "changes his knife every month", because his knife, even if initially of good quality, might get damaged when encountering different circumstances (various oxen). A good cook, however, possesses a greater grasp of the *Dao* of Butchering than an ordinary cook, enabling him to adapt to more circumstances than an ordinary cook. This translates to less frequent knife replacements—"every year" compared to "every month".

Zhuangzi not only praises *Dao* over artifacts in the story of butcher Pao Ding. Moreover, another tale in *Zhuangzi* further exemplifies Zhuangzi's skepticism towards techniques. In the tale, when Confucius's disciple Zigong encounters an old man manually carrying water from a well to his farm, he suggests a more efficient contrivance for irrigation. The gardener, however, dismisses the idea, expressing a preference for manual labor despite its inefficiency, and remarks that

> I have heard from my teacher that, where there are ingenious contrivances, there are sure to be subtle doings; and that where there are subtle doings, a scheming mind is surely at play. However, such a scheming mind within one's heart taints its inherent purity. This tarnished simplicity disrupts the spirit, rendering it an unfit *vessel* for the *Dao*. It's not that I'm unaware (of the contrivance you speak of), but I would find it dishonorable to employ it (Zhuangzi, "Heaven and Earth", 2006).

For Zhuangzi, the pursuit of ingenious contrivances leads to a scheming mind, disrupting the purity of the soul and unsettling the spirit, which is not conducive to *Dao*.

In summary, despite the variance in the *Dao-Qi* relation between Confucius and Zhuangzi, a commonality they share is prioritizing *Dao* over *Qi*. Confucius emphasizes the importance of tools, while also suggesting that a gentleman should not be bound to specific purposes like artifacts. Zhuangzi, however, exhibits a much stronger skepticism towards technology, for he views artifacts not only as limited by circumstance but also inherently harboring a "scheming mind" within them, disrupting the spirit and corrupting the user's connection with the *Dao*.

2.4 Legalists' Attitude to Techniques: *The Book of Lord Shang*

It appears that Confucianism harbored some implicit and mild hostility towards technology, while Taoism was more explicit in their criticism of it. However, compared to both, Legalism's attitude to techniques appears harshest.

The Book of Lord Shang, a classic text of the School of Legalism, documents the policies advised by Lord Shang, Shangyang (390–338 B.C.), to the ruler of the State of Qin during the Warring States Period (453–221 B.C.), when conflicts among states were prevalent. Lord Shang, a contemporary of Zhuangzi, was tasked with implementing a series of reforms in the State of Qin, laying down administrative, political, and economic foundations that eventually enabled the State of Qin to conquer the other six rival states, unifying China under centralized rule for the first time in history. The policies advocated by Lord Shang in The *Book of Lord Shang* express explicit and strong hostility towards technology.

> Nowadays, people within the territory all believe that by avoiding agriculture and war, one can attain office and rank. Consequently…insignificant individuals will occupy themselves with trade and will practise crafts, all in order to avoid agriculture and war, thereby creating a perilous situation for the state. When the populace subscribes to such teachings, it is inevitable that such a country will disintegrate…. If there are a hundred people engaged in agriculture and war and only one in the crafts, then those hundred will all be remiss in agriculture and war (Shang Yang, 2022).

The above passage illustrates Legalism's apparent disdain for crafts and especially their practitioners. Unlike Confucius and Zhuangzi, whose "critiques" of technology lie in its impact on *Dao-Qi* relation or its ethical implications (tainting inherent purity of heart, disrupting the spirit, etc.), Lord Shang focuses on the social and political consequences of technology. However, his concern isn't with technology as a transformative force for society or politics, as argued by modern thinkers like Marcuse (2006) or Feenberg (2002). Instead, Lord Shang worries about the socio-political implications of making artifacts as a lifestyle or a way of making a living.

Once craftsman becomes a profession in society, they are inevitably fraught with socio-political consequences. From the Legalist perspective, agriculture and war are paramount to a nation's power then. Embracing crafts as an occupation would lead to a decreased willingness to engage in farming or warfare, thus undermining the state's power. In Lord Shang's time, an era of frequent wars among states, decrease of a state's power sometimes could be an existential threat to it.

It's important to clarify a potential misconception: Legalism does not explicitly express inherent animosity towards crafts themselves. Rather, it disapproves of crafting as a life pursuit. However, Legalism's significant influence on the Qin state is undeniable. Lord Shang led two Legalism reforms in the State of Qin and served as chancellor there for a decade (Sima Qian, c. 91 B.C./2022). His clear disdain for craftsmen, in turn, undoubtedly left a lasting impact on societal attitudes towards crafts during that period.

However, even if a Legalist society discourages crafts as a lifestyle, it would still require skilled artisans to produce necessary tools and make technological advancements. So, the question arises: who would fulfill this crucial role? One potential source of these craftsmen could be government-sponsored sections dedicated to specific crafts.

2.5 Techniques and Artifacts as the Manifestation the Wisdom of Sages: *The Rites of Zhou*

Across the intellectual landscape from Confucius (551–479 B.C.) to Zhuangzi (369–286 B.C.) and Lord Shang (390–338 B.C.), a growing skepticism towards technology seems to emerge, culminating even in hostility. Confucius viewed artifacts as tools with defined purposes, and their existence status represented a state that people should strive to avoid. Zhuangzi, in contrast, emphasized the *Dao* as the key to adaptability. He saw artifacts as limited to specific uses, efficient within those parameters but ultimately limiting one's ability to navigate diverse situations. Furthermore, Zhuangzi believed artifacts inherently embodied a "scheming mind," and the pursuit and reliance on them would cultivate such a mindset, ultimately disrupting the purity of the soul and unsettling the spirit. Lord Shang, however, focused on the social and political consequences of crafting as a profession, believing it would undermine a state's power.

Confucius lived in the Spring and Autumn Period (770–453 B.C.), while Zhuangzi and Lord Shang lived in the Warring States Period (453–221 B.C.) (Sima Guang et al., 1084/2011). This raises a question: Does the growing skepticism towards techniques from Confucius to Zhuangzi and Lord Shang reflect a coincidence, or is it part of a larger historical trend? If we delve back in time to the period before Confucius, would we find a more positive attitude towards techniques (Table 2.1)?

Examining records on techniques and craftsmen from the pre-Spring and Autumn Period to the early period of Zhou Dynasty (1046–256 B.C.) reveals a contrasting attitude to techniques. *The Rites of Zhou* (周礼), a book recording rituals in the early Zhou Dynasty period (c.1042–c.1035 B.C.),[3] provides a glimpse. According to *The Rites of Zhou*, the early Zhou Dynasty society was divided into six professions: the aristocracy, scholar-officials, various craftsmen, merchants, peasants, and seamstresses. Apparently, comparing to Legalism society, the inclusion of craftsmen as a distinct social class suggests a higher regard for techniques and craftsmen in the early Zhou Dynasty.

[3] There are debates on the author of *The Rites of Zhou*. However, the book itself delves into lists outlining the bureaucratic structure of the early Zhou Dynasty.

Table 2.1 Shifts in attitudes towards technology in traditional Chinese thought

Period	People	Attitude towards technology	Changes in attitudes	
Before The Spring Autumn Period	?	?	–	From bottom to top↑: Will there be an increasingly welcoming attitude towards technology?
The Spring Autumn Period (770–453 B.C.)	Confucius (551–479 B.C.)	Artifacts are essential for fulfilling specific purposes, yet they are inherently limited to specific purposes	From top to bottom↓: There is an increasing skepticism, and even intense hostility, towards technology	
The Warring States Period (453–221 B.C.)	Zhuangzi (369–286 B.C.)	Artifacts are inferior to the *Dao*, and inherently embodied a "scheming mind"		
	Lord Shang (390–338 B.C.)	Crafting as a life pursuit would undermine a state's power and is disdained		

In *The Rites of Zhou*, craftsmanship is not only revered as a principal vocation alongside the aristocracy and scholar-officials, but is also considered a domain associated with the Sages.

> It is the Sages who innovate, while artisans inherit and preserve these creations, earning them the title of craftsmen in society. The works of countless craftsmen are, fundamentally, *extensions of the Sages' ingenuity*. They mold metal into blades, shape clay into vessels, construct carts for terrestrial travel, and build boats for aquatic navigation—all stemming from the brilliance of the Sages (The Duke of Zhou, 2022).

What these esteemed Sages introduce transcends mere technical skills. They offer more sophisticated ideas, emphasizing *a harmonious relationship between heaven, earth, materials, and the craftsmen.*

> Heaven operates in its seasons, the earth harbors its vital energy (气, *qi*), materials embody their intrinsic beauty, and craftsmen wield their expertise. It is the convergence of these four parties together that gives birth to an exquisite creation. If the materials are of superior quality and the craftsmanship exceptional, yet the outcome falls short, it must be attributed to either an unfavorable timing or a discord with the earth's vital energy (The Duke of Zhou, 2022).

Nonetheless, the Sages invent artifacts into the world, and it is the craftsmen who follow in their footsteps, replicating and preserving these innovations. In Chinese culture, the term "Sage" represents the highest accolade one can bestow upon an individual. The artifacts crafted are a testament to the profound wisdom of these revered Sages. Craftsmen serve as both practitioners and custodians of the Sages'

Table 2.2 A growing skepticism towards technology in Pre-Qin Chinese thought

Early Zhou Dynasty	*The Rites of Zhou* (c.1042–c.1035 B.C.)	An increasing skepticism, and even intense hostility, towards technology from top to bottom ↓
Spring and Autumn Period	Confucius (551–479 B.C.)	
Warring States Periods	Zhuangzi (369–286 B.C.)	
	Lord Shang (390–338 B.C.)	

legacy. Clearly, during the early Western Zhou Dynasty, craftsmen enjoyed high esteem.

Examinations across Sects. 2.2–2.5 reveals shifts in the attitudes towards technology in traditional Chinese technological understanding (Table 2.2). In the early period of Zhou Dynasty (c.1042–c.1035 B.C.), as depicted in *The Rites of Zhou*, techniques were highly revered. They were seen as manifestations of the Sages' wisdom, embodying a harmonious relationship between heaven, earth, materials, and the craftsmen themselves. However, by the time of Confucius (551–479 B.C.) during the Spring and Autumn Period, this reverence began to wane. Techniques were still viewed as *necessary* tools by Confucius, but their connection to the divine and the ideal of harmony was diminished. Confucius saw them as purely functional means to specific ends. This shift continued during the Warring States Period (475–221 B.C.) within the philosophies of figures like Lord Yang (390–338 B.C.) and Zhuangzi (369–286 B.C.). In *Zhuangzhi*, the story of Butcher Pao Ding exemplifies a partial detachment of techniques from the *Dao*, the underlying order of the universe, and moreover, reliance on technology, it was argued, could disrupt the purity of the soul and unsettle the spirit. Furthermore, Shang Yang even saw craftsmen as a threat to state power, potentially even an existential risk in the age of constant warfare. Throughout this progression or regression, reverence for techniques and craftsmen waned while skepticism towards them continued to mount.

2.6 Socio-Political Aspects of the Shifts

As we observe a more open acceptance of techniques in the period preceding the Spring and Autumn Period compared to the Spring and Autumn Period and the Warring States period, a trend seems to emerge. It becomes increasingly likely that the growing skepticism towards these techniques, from Confucius to Zhuangzi and Lord Shang, was not merely coincidental. If this skepticism did indeed reflect a larger historical shift, what factors might have contributed? One plausible explanation for the evolving attitudes of the ancient Chinese elite towards techniques and the artisan class might lie in the changing socio-political landscape across the centuries following the establishment of the Zhou Dynasty.

The Zhou Dynasty thrived under the Duke of Zhou's leadership (c.1042–c.1035 B.C.). He became a paragon of rulership in traditional Chinese culture, credited with ushering in an era of peace and order. Confucius, reflecting on this period in his writings, held the early Western Zhou Dynasty in high esteem. This admiration extended to "*The Rites of Zhou*," a classic text Confucius revered as an exemplar of the ideal social structure. "The Master said, 'Zhou had the advantage of the two past dynasties. How complete and elegant are its regulations! I follow Zhou'" (Confucius, 2006). By the time of Confucius (551–479 B.C.), Chinese society had descended into greater chaos compared to the peaceful days in the early Western Zhou.[4] Nobles were warring with each other, showing scant respect for the rulers of the Zhou Dynasty, who were nominally still the Sons of Heaven. The lower ranks of the aristocracy were overthrowing their superiors in a series of power struggles, accompanied by warfare that spilled over into the peasantry (Chen, 2018). This upheaval ran counter to the harmony advocated by Confucian teachings. During the era of Shang Yang (390–338 B.C.) and Zhuangzi (369–286 B.C.), approximately a century and half after Confucius, societal chaos in China intensified, and conflicts among nobles and kingdoms became more and more frequent (Chen, 2018).

The escalating skepticism towards technology, as shown in Table 2.2, may well be correlated with the significant decline in social and political order during the transition from the early Zhou Dynasty to the Warring States Period. *As warfare became increasingly frequent, a societal shift towards pragmatic aspects is understandable*. Consequently, instrumental aspects of techniques gains more and more interests in technological understanding. Similarly, agriculture and military affairs which were seen as "utensils" particularly for the rulers of the kingdoms, for the increase of power, gained heightened importance. Rulers sought to encourage more individuals to engage in farming and warfare, leading to a diminishing emphasis on the artisan class, as evidenced in *The Book of Lord Shang*.

Not only the socio-political atmosphere changes influenced people technological understanding, the shifts in technological understanding also deep influenced the fate of techniques, social ranks of craftsmen and even the standing of the Ministry of Works in ancient China. The perception of technology as utensils meant that its value was measured solely by its utility to rulers and their "grand causes". If technological advancements did not directly serve the interests of the ruling elites, they were often neglected or even actively suppressed. Consequently, the social status of craftsmen suffered, as evidenced by historical accounts such as *the Book of Lord Shang*.

Moreover, historical research reveals that the amount of officials in the Ministry of Works was often relatively small across most dynasties in China after Zhou Dynasty. Additionally, scholar-officials who worked within the Ministry of Works typically did not enjoy favorable prospects for political advancement (Wei, 2010; Yang, 2013).

[4] The Zhou Dynasty (1046–256 B.C.), due to a war, relocated its capital eastward in 770 B.C. This shift marked a turning point, with the western capital period known as the Western Zhou and the subsequent eastern capital period referred to as the Eastern Zhou.

2.7 Conclusion

Except the well-recognized Confucian and Taoists' technological thought, there is also a dimension of technological thought in Legalism, especially in *the Book of Lord Shang* by Shangyang. By delving into the often-overlooked dimensions of technological thought within Legalism and contrasting them with the technological perspectives found in *The Rites of Zhou*, the teachings of Confucius, and the philosophy of Zhuangzi, we can see the shifts of technological understanding in pre-Qin Chinese thought. The evolution in technological thought precipitated significant changes in attitudes towards techniques and the artisan class. These marked changes in thought and attitude are not only tied to the tumultuous transformations in the social-political landscape of the era but also carry profound implications for the socio-political fabric of traditional Chinese culture.

References

Chen, S. (2018). The age of territorial lords. In P. Goldin (Ed.), *Routledge handbook of early Chinese history* (pp. 108–145). Routledge.

Confucius. (2006). *The Analects*. Zhonghua Book Company.

Feenberg, A. (2002). *Transforming technology: A critical theory revisited*. Oxford University Press.

Hui, Y. (2016). *The question concerning technology in China: An essay in cosmotechnics*. Urbanomic.

Legge, J. (Ed.). (2009). *Confucian Analects, the great learning & the doctrine of the mean*. Cosimo Classics.

Marcuse, H. (2006). *One-dimensional man*. Routledge.

Shang Yang. (2022). *The book of Lord Shang*. Zhonghua Book Company.

Sima Guang, et al. (1084/2011). *Zizhi Tongjian*. Zhonghua Book Company.

Sima Qian. (c. 91 B.C./2022). Biography of Lord Shang. In *Records of the grand historian*. Zhonghua Book Company.

Tai, Z. (2008). *The history of Chinese philosophy*, Introduction. East Press.

The Duke of Zhou. (2022). *The Rites of Zhou*. Zhonghua Book Company.

Wei, L. (2010). *The research on Gong Bu Shang Shu in Tang Dynasty*. Doctoral Dissertation, Shangdong University.

Yang, J. G. (2013). *The research on Gong Bu in Song Dynasty*. Doctoral Dissertation, Shanghai Normal University.

Yang, T. C., & Zhang, S. W. (Trans.) (2011). *The book of changes*. Zhonghua Book Company.

Zhu Xi. (2012). *Commentaries on the Analects*. Zhejiang University Press.

Zhuangzi. (2006). *Zhuangzi*. Zhonghua Book Company.

Dong Luo is an Associate Professor at the Institute for Advanced Studies in Philosophy, Science, and Technology at South China University of Technology, P.R.C. One of his recent research focuses is the interplay between scientific and technological understanding within social and political contexts. He is the General Secretary of the Chinese Association of Philosophy of Physics.

Chapter 3
A Constructively Positive View on Technology: A Response to Carl Mitcham's Question

Tom Xiaowei Wang

Abstract This paper responds to Carl Mitcham's critical reflections on the philosophy of technology in China, particularly his observation of its generally positive stance toward technological development. While acknowledging this optimistic tone, we argue that Chinese scholarship harbors a more nuanced and evolving view, shaped by Marxist theory, post-phenomenology, and emerging Confucian perspectives. Through a historical and philosophical review, this chapter traces the development of the philosophy of technology in China, from early conflations of science and technology to current efforts to construct native theoretical frameworks. Special attention is given to the potential of a Confucian philosophy of technology as a culturally grounded, constructively critical approach that balances ethical concerns, societal harmony, and technological innovation.

Keywords Philosophy of technology · Carl Mitcham · Confucianism · Marxism · Post-phenomenology

3.1 Background

With his deep love for China and genuine concern about the Chinese academia in the philosophy of technology, Professor Carl Mitcham has expressed worry about how the philosophy of technology is studied in China. Mitcham authored a paper reflecting on the development of the philosophy of technology in China, which sparked widespread responses from both emerging and established Chinese scholars. This vibrant discussion culminated in the publication of a dedicated volume in *The Development of Philosophy in China* (哲学动态), one of China's leading philosophy journals. Contributors to this volume addressed a range of critical issues

T. X. Wang (✉)
Renmin University of China, Beijing, China

L. Checketts (ed.), *What Went Wrong with the West: Carl Mitcham's Critique of Western Philosophy of Technology*, Philosophy of Engineering and Technology 52, https://doi.org/10.1007/978-3-032-22214-5_3

raised by Mitcham's work, exploring topics such as the need for a critical approach in the philosophy of technology, the distinction between "big questions" and "small questions," and the future direction of Chinese philosophy of technology, including its internationalization and governance.[1]

In his contribution to the volume, Carl Mitcham has observed that Chinese scholars tend to share a generally positive view of technology, a perspective similar to that of many American philosophers. The work of Don Ihde (2008, 2012), whose theories focus on specific, small-scale questions about technology, has been widely translated and well-received in China. This contrasts with the more critical approaches to the philosophy of technology taken by scholars like Professor Andrew Feenberg and Albert Borgmann. In particular, the works of Ivan Illich, the mentor of Professor Mitcham, have largely remained untranslated. To date, only one of Illich's books, *Deschooling Society* (2017), has been translated into Chinese, leaving his more critical perspectives on technology relatively unknown in China. For Carl Mitcham, the work of Ivan Illich delves deeper into the philosophical implications of the widespread adoption of modern technology as a system, rather than focusing on individual artifacts.

Illich's work spans a wide range of topics, including but not limited to medical technology, education as a technological system, tools as a mode of existence, and gender. His ideas have had a significant impact on American society since the 1970s, yet they remain largely unknown to Chinese audiences. Illich takes a critical stance on the adoption of technology, viewing it not as a collection of individual instruments that we can use at will, but as an encompassing system in which we all live. Technology, in his view, represents power structures and functions as a rationale that coordinates human coexistence. Thus, it is crucial to recognize the political nature of technology and to examine its power dynamics. I believe that Mitcham aims to encourage Chinese scholars, particularly the younger generation, to analyze technology from this unique perspective, enabling a deeper understanding of the complexities inherent in modern technologies.

In this chapter, I will attempt to address Mitcham's question of why Chinese scholars tend to hold a generally positive view of technology. My argument is that despite this seemingly positive outlook, Chinese scholars actually hold a more nuanced view of technology than it may appear. I will briefly introduce how the field of the philosophy of technology has developed, and then provide a narrative explaining how a more critical discourse of philosophy of technology in China has been developing.

[1] See the articles in *The Development of Philosophy in China*, 2021, (01): 25–52. 哲学动态, 2021, (01): 25–52.

3.2 When Science Departs from Technology

For a long time, the terms "science" and "technology," which are distinct in English, were often conflated into a single word in Chinese, namely "sci-tech" (科技). Contemporary Chinese understanding of science is often framed through the lens of technology. During the late Qing dynasty, China encountered Western powers through tangible artifacts such as ironclad ships and heavy cannons, which were perceived as the embodiment of scientific advancement. The Qing government urged prominent figures to introduce Western technologies, particularly military technologies, to strengthen national security.

After establishment of the People's Republic of China, China developed a long-term science and technology development plan for the period from 1956 to 1967, and 2 years later, the Chinese National Science and Technology Council was established. The term "sci-tech" gradually became part of everyday language. Even today, the distinction between science and technology remains blurred, as they are often used interchangeably. However, more and more philosophers have begun to recognize the fundamental differences between these two concepts. In 1982, Chen Changshu published the article "The Unity and Differences of Science and Technology"(科学与技术的统一与差异) in *Guangming Newspaper*, which distinguished between the concepts of science and technology.

In his recent book *What Is Science?* (什么是科学), Professor Wu Gusheng (2016) explains that technology is a form of practical wisdom. Across all civilizations, people have invented various tools and artifacts to solve their tasks. Science, however, is a different matter; it was first conceptualized by the ancient Greeks as a distinct endeavor for pursuing humanity. For the ancient Greeks, science was not about technological application but about understanding the universal, immutable laws of the cosmos. A person capable of appreciating geometry and music was considered cultured.

Since the 1980s, the philosophy of science has garnered significant attention in China, whereas the philosophy of technology has been studied by only a few. However, the research landscape is undergoing a significant transformation. More and more scholars have recognized the importance of studying the philosophy of technology.[2] Today, science has evolved into techno-science; no scientific progress can be made without the use of various technological instruments. With advancements in artificial intelligence and biotechnology, it has become urgent for scholars to understand the essence of the mind and human nature. Technology is no longer widely seen as a neutral instrument to be used for whatever purposes humans desire. Instead, it is increasingly viewed as a mediator that constitutes fundamental aspects of human life.

---

[2] Recently, some prominent Chinese philosophers, such as Zhao Tingyang (赵汀阳) (2018) and Jiang Yi (江怡) (2023), have begun publishing articles on AI philosophy and engineering philosophy.

Given this context, many philosophers are now advocating for a shift from the philosophy of science to the philosophy of technology, and an increasing number of Chinese scholars are beginning to take the philosophy of technology seriously, recognizing that it has become a mainstream topic in contemporary philosophical studies. Scholars who have not traditionally worked in the departments of philosophy of science and technology are also beginning to explore philosophical issues related to technology from their own unique backgrounds.

3.3 A Brief History of Philosophy of Technology in China

As of 2024, there have been 4071 papers published on the philosophy of technology. The number of publications has increased significantly since 2000, with an average of around 200 papers being published each year now. Below, I have listed some essential literature to give readers a general sense of how the philosophy of technology, as a field of study, has developed in China:

The earliest documented reference to the philosophy of technology in China can be traced back to a report in *Philosophical Translations* (哲学译丛) in 1978, which discussed a conference on the philosophical issues of technical sciences held in East Germany (Zao, 1978; Serezoe & Wu, 1978). In 1980, the *Philosophical Translations* series introduces the philosophy of technology in the Americas (美洲的技术哲学<哲学译丛) including the works of George Grant, Henryk Skolimowski, Hans Jonas, and William Leiss.

Chen Changshu (July 1932—March 20, 2011), widely seen as one of the pioneering philosophers of technology in China, served as a professor and director of Northeastern University of China's School of Humanities and Law. He was a member of the Philosophy Review Panel for the second, third and fourth State Council Academic Degrees Committees and Vice President of the Chinese Society for Dialectics of Nature. Chen graduated from Northeastern University of China's Mining Department in 1954 and completed his graduate studies in philosophy at Renmin University in 1956. In 1983, Chen published a paper titled, "A Brief Discussion on the Study of the Philosophy of Technology" (简论技术哲学的研究), was published in the *Journal of Northeastern Institute of Technology* (东北工学院学报). In 1985, Chen authored "Technology as a Subject of Philosophical Inquiry" (技术是哲学的研究对象), published in *Studies in Dialectics of Nature* (自然辩证法研究). In 1986, Chen published "The Influence of Technology on the Development of Philosophy" (技术对哲学发展的影响) in the same journal, and later presented his landmark report, "The Need to Develop the Study of the Philosophy of Technology" (要开展对技术哲学的研究), at the Second National Congress of the Chinese Society for Dialectics of Nature. Since then, the philosophy of technology has grown.

Generally speaking, in Chinese academia, the philosophy of technology is approached from three main directions. The first direction is rooted in the analytical tradition. Many of these scholars, often from the Department of Western Philosophy, have begun to clarify complex concepts. For instance, they grapple with questions such as: What is the essence of mind? What are qualia? What is nature of technology? What are the differences between various approaches to design artificial intelligence? Which approach is more effective or efficient? These inquiries often overlap with the work of computer scientists and neuroscientists, as philosophers in this tradition hope that their conceptual clarifications might sometimes guide scientific research.

The second approach comes from the Science, Technology, and Society (STS) perspective. Scholars in ethics, applied ethics, and departments of Science and Technology have begun to investigate how technology interacts with various societal elements, such as politics and moral norms, which can ultimately transform our ethical practices. Due to the open and inclusive nature of STS, this line of research has attracted a diverse range of talents, not only from ethics and the philosophy of science and technology but also from fields such as political science and sociology. However, there is growing concern within the academic community that the boundaries of STS have become increasingly blurred, with worries that scholars are drifting away from traditional philosophical inquiry and encroaching on the domain of sociology. This approach, while rich in empirical data, risks becoming overly empirically driven, potentially overshadowing the distinct methodologies that are central to philosophical analysis.

The third approach to the philosophy of technology in China is the phenomenological perspective. Chinese philosophers, particularly those from the older generation born in the 1960s, have been deeply influenced by the work of Martin Heidegger. Heidegger's philosophy of technology was introduced to China in 1992 when Mo Weimin and Gao Lianghua published articles in "Exploration and Free Views" (探索与争鸣) and the *Journal of Dialectics of Nature* (自然辩证法通讯) titled "An Analysis of Heidegger's Philosophy of Technology" (试析海德格尔的技术哲学) and "On Heidegger's Philosophy of Technology" (论海德格尔的技术哲学), respectively. Heidegger's critique of technology is a key framework they often use to understand the essence and impact of technology.

In recent years, the post-phenomenological perspective on technology has gained increasing popularity among younger scholars. Don Ihde's work was introduced in 1994, with the article "Reflecting on the Meaning of Technology in the 'Lifeworld': An Introduction to Don Ihde's Philosophy of Technology" (在"生活世界"里反思技术的意义——伊德技术哲学初探) (Zhang, 1994). Since 2008, the Dutch school of philosophy of technology—exemplified by Peter-Paul Verbeek's work on the moralization of technology and Peter Kroes's research on the dual nature of artifacts—has gained increasing attention among younger scholars in China, extending its influence to fields such as media studies and industrial design.

Unlike traditional phenomenology, which focuses on the grand narratives and ontological preconditions of technology, post-phenomenology shifts attention to individual artifacts and their impact on our daily lives. This approach is proving to be particularly fruitful for understanding our contemporary technological situation, where everyday life is increasingly mediated by a vast array of technologies.

Technology permeates every aspect of our existence, influencing how we experience the world.

This brief historical review identifies three primary influences on the study of the philosophy of technology in China: The analytical tradition pays little attention to evaluating technology; instead, it focuses on explaining it. Closely following the most recent technological advancements, it often appears to be pro-technology in the eyes of many. STS approaches often focus on case-by-case analyses rather than broader philosophical narratives. Consequently, they tend to adopt a more positive view of technology, given that many widely adopted artifacts are perceived as having a beneficial impact on society.

More importantly, from STS perspective, one should note that the institutionalization of the study of the philosophy of technology in China has occurred within the framework of the Chinese Societies for the Study of Dialectics of Nature, a branch of knowledge initiated by Engels as part of Marxist thought. This society was established to bring together scholars from both scientific and humanities backgrounds to explore interdisciplinary questions. The Society for the Philosophy of Technology, in particular, was created to deepen understanding of science and technology, thereby promoting their development.

Within the core of Marxism, science and technology are often viewed as instruments for driving productivity (Liu, 1992). Capitalists can exploit these tools to increase efficiency and accumulate capital. However, this dynamic can also lead to collusion between scientists, technologists, and capitalists, creating labor conditions in which the proletariat suffers. Nevertheless, among Chinese Marxist scholars, there is a consensus that Marxism views science and technology as class-neutral. These tools are not inherently tied to any specific class; they can be utilized by both capitalists and the proletariat (He et al., 1978). Deng Xiaoping emphasized that science and technology are the primary forces of productivity (Yang, 1991). Consequently, they are regarded as neutral instruments capable of significantly increasing productivity, thereby boosting social wealth in preparation for a future communist society.

Moreover, Marxist philosophy emphasizes making concrete changes and highlights the agency of human beings. Within this framework, science and technology are recognized as possessing transformative power, but they are seen as tools that can and should be controlled by human agents. The central issue is who should control technology. Marxists argue that if the proletariat can harness and control technology, it can be used to create a just society. This Marxist perspective on technology has fundamentally shaped Chinese studies in the philosophy of technology. This view, widely shared among Chinese scholars, contributes to a generally positive interpretation of Marxism's stance on technology.

In contrast, thinkers such as Karl Jaspers, Jacques Ellul, Albert Borgmann, and Martin Heidegger present a different view of the relationship between technology and humanity. These scholars argue that technology is not merely an instrument controlled by humans. Instead, they view technology as a mode of existence or an autonomous system that cannot be fully controlled by humans. This critique is very different from the human-centered perspective of Marxism. Heidegger's critique of technology has garnered much attention from Chinese scholars. Heidegger's critique holds great significance as it elevates the philosophy of technology to the level of first philosophy, specifically ontology.

However, while Heidegger's ontological analysis provides profound insights, it is not as effective when it comes to addressing how specific, concrete artifacts impact our daily lives. As a result, younger scholars increasingly turn to STS and post-phenomenological perspectives. As such, the general atmosphere surrounding Heidegger's grand narrative critique of technology is somewhat confined to a small circle of scholars, particularly among the older generation. These scholars tend to be more critical of technology, but they remain a minority. In the broader Chinese context, where science and technology are often conflated as synonymous with positive productivity growth, Heidegger's critiques are frequently dismissed as overly romantic or even pessimistic.

The post-phenomenological approach occupies a middle ground between ontological investigation and applied ethics. It provides insights into the fundamental aspects of technology while remaining properly informed by empirical data. This balanced perspective helps bridge the gap between abstract philosophical inquiry and the practical implications of technology. When conducting a literature review, one cannot help but be struck by the sheer volume of publications devoted to studying and introducing post-phenomenological perspectives on technology, particularly in the context of moralizing technology.

Current research has progressed to the point where these perspectives are being applied to guide both artistic and industrial practices. Here, the philosophy of technology is broadly regarded as a valuable resource that can be harnessed to promote more responsible technological practices. Ethics, in this context, is not viewed as an external force aimed at curbing the rapid development of science and technology; rather, it is seen as a constitutive element in the creation of better products.

This shift in perspective has made the philosophy—and particularly the ethics—of technology directly relevant to contemporary technological practices. In China, this view has been embraced by many technologists and engineers. However, this trend also raises concerns. One could argue that this shift signifies the loss of philosophy's distinct status as something independent of technology. Where once the role of ethics and philosophy was to supervise and reflect on technological advancements, they now risk becoming subordinate to technological rationality. Philosophy, in this framework, is reduced to elements that need to be embedded within technology to make it more palatable to the general public. Professor Carl Mitcham, however, is notably dissatisfied with this generally positive attitude, whether explicit or implicit, towards technology. At the same time, he finds the ontological critique of technology lacking in specificity and practical application, as it is often too abstract to drive meaningful societal change.

3.4 Imagining a Critical Theory of the Philosophy of Technology

One way to expand the critical study of the philosophy of technology is to translate more Western works into Chinese. For instance, Ellul's book on technological systems remains untranslated and would be valuable. Bernard Stiegler's work has been translated but hasn't garnered much attention in China. The French school of philosophy, including thinkers like Simondon, offers a nuanced view of technology as central to human existence, not just as a tool. Works by Albert Borgmann, Andrew Feenberg, and especially Ivan Illich, whose books are largely untranslated into Chinese, should also be introduced to Chinese scholars.

Nevertheless, we are skeptical about the effectiveness of this approach. Unlike classical studies, where Chinese and Western scholars began research at different times, the philosophy of technology developed almost simultaneously in both regions. Therefore, there is no established "canon" in this field comparable to the German or Greek classics. Additionally, younger Chinese scholars are increasingly less interested in translating Western texts, as many are trained abroad and already proficient in French, German, and English. They often admire these works but are reluctant to undertake translations themselves.

On the other hand, the Chinese government is encouraging scholars to build a knowledge system that integrates Marxist thought and traditional Chinese culture. This aims to create a unique framework for engaging with Western theories from a Chinese perspective. Given this context, simply translating Western works may not be enough to develop a more nuanced understanding of technology. We are not suggesting that translation of Western works should cease; I believe that translation remains valuable. However, my concern is that merely increasing the number of translations may not necessarily lead to the more reflective attitude towards technology that Professor Carl Mitcham envisioned.

Introducing Western theories provides valuable perspectives but does not necessarily generate substantive research questions. Real research questions emerge from specific contexts. China's experience with technology since the late Qing Dynasty offers a unique narrative where technology is seen as a powerful force for transforming reality. Marxism views technology as a force representing an advanced productivity. These narratives and historical contexts are crucial, as they shape our interpretative framework. To develop a comprehensive Chinese narrative on technology, we must learn from Western perspectives while focusing on forming research questions grounded in our own historical and cultural contexts.

One feasible approach is to explore Marxist studies more deeply, focusing on how Marx and Engels discussed the benefits and drawbacks of technology. In Marxism, technology is analyzed within its political and economic contexts, rather than in isolation. It is seen as interacting with social relationships and factors, shaping and being shaped by them. We need to examine how technology impacts factory work and transforms the nature of labor, fundamentally altering social productive relationships. For example, many scholars are now investigating digital labor.

Today, work often extends beyond traditional factory settings, with many people working from home or engaging in activities that were previously not considered work. Activities like using TikTok, once seen as leisure, now generate data that companies use for targeted advertising. Marxist political economy remains relevant for contemporary discussions on the philosophy of technology. It offers a nuanced framework for understanding how technology influences and is influenced by social and economic conditions.

In fact, there has been a notable increase in papers published on the philosophy of technology from a Marxist perspective. However, current discussions largely focus on specific topics, such as applying Marxist notions of labor to new phenomena like digital technologies. Despite these efforts, there has been limited progress in developing a comprehensive Marxist philosophy of technology. We are confident that younger scholars will increasingly take on the task of collecting and analyzing Marx's and Engels' work on machines to develop a comprehensive Marxist theory of technology.

Another approach I find particularly fascinating is the integration of traditional Chinese philosophies, such as Confucianism into the study of technology. This involves more than merely applying Confucian concepts to technological challenges, as is commonly done in fields like robotics ethics, engineering ethics, and bioethics. While existing work often explores how Confucian ideas about life influence ethical principles, such as informed consent, this is generally categorized under applied ethics.

A true Confucian philosophy of technology would be distinct. It should involve using traditional Chinese wisdom to develop a broader theoretical framework for understanding technology, rather than just applying these concepts to specific ethical issues. In 2005, Wang Qian published "From Skill to Dao: The Philosophy of Technology in Chinese Tradition" (由技至道——中国传统的技术哲学理念). Li Sanhu published another such project, the book *Reaffirming Tradition: A Holistic Comparative Study of the Philosophy of Technology* (重申传统:一种整体论的比较技术哲学研究) in 2009. These two works are exceptionally original in establishing a grand narrative of philosophy of technology grounded from Chinese ancient traditions.

Heavily influenced by these two authors, Yuk Hui (2019) published his monograph *The Question Concerning Technology in China: An Essay in Cosmotechnics*, which has gained international recognition. Pak-Hang Wong and I (2021) also contributed to the field by publishing the first edited volume dedicated to exploring Confucian ethics in technology, titled *Harmonious Technology: Confucian Ethics of Technology*. This volume brings together perspectives from scholars who discuss the ethics of technology through the lenses of post-phenomenology, the history of science and technology, engineering ethics, etc.

Currently, there is a consensus on the need to diversify the discourse in the philosophy of technology. This is an urgent task, especially since the philosophy of technology, compared to the philosophy of science, is more practice-oriented. For instance, discussions on the ethics of artificial intelligence often focus on AI governance, with ethics and philosophy playing key roles in shaping policy. As technology operates on a global scale, AI and related applications, once released, are quickly adopted by people worldwide. A critical philosophy of technology is not an abstract or ivory tower practice; it is grounded in politics, economics, and everyday life. Given this, there is a growing need to explore and construct a non-Western narrative in the philosophy of technology. To achieve this, we believe three key aspects must be addressed.

First, constructing a Chinese philosophy of technology requires a comparative perspective. Traditional Chinese civilization did not have a concept equivalent to the philosophy or ethics of technology, and the very term "philosophy" is foreign to Chinese culture, originating from ancient Greece. Even in the West, discussing the philosophical implications of technology is a relatively modern endeavor. Therefore, when looking back at Chinese traditions, it is essential to approach them with a solid understanding of Western philosophy of technology. This inherently comparative approach means we are not merely reviving sporadic discussions of technology in Chinese classics. Instead, we aim to reconstruct and reconsider the Chinese tradition to develop a more comprehensive framework for understanding technology.

Second, our research questions should be deeply rooted in both the global and Chinese contexts. While we can learn from Western methodologies in philosophy and specifically in the philosophy of technology—drawing on thinkers like Heidegger, Ellul, Borgmann, and others—our questions must be grounded in China with a global perspective. We should explore how technology was perceived in ancient China, how technological elements interacted with other societal factors such as ethics and politics, and how it contributed to societal cohesion, and this inquiry is particularly relevant when considering how attitudes toward technology evolved from the late Qing dynasty, when China first opened to the Western world, through to the establishment of the People's Republic of China in 1949, and beyond.

Third, we need to examine how science is understood through the lens of technology and how technology, initially perceived as an external, foreign power, gradually became integrated into Chinese and socialist rationality. Additionally, we must delve into Chinese traditions—whether Daoism, Confucianism, or Buddhism—to see how these cultural frameworks provide resources for understanding technology and technological artifacts. From an ontological perspective, we must explore the metaphysical conditions that made technology conceivable and tangible. we must investigate the fundamental elements that condition technology, making it visible and tangible to the average person.

For example, Heidegger, in discussing the essence of technology, begins with instrumentality and then draws on Aristotle's theory of the four causes to uncover the deeper metaphysical conditions that make technology possible. Similarly, we need to delve into ancient Chinese traditions to explore comparable foundational elements. In addition, we need to investigate how specific technological artifacts were understood in relation to other societal factors. It is essential to examine how technological artifacts—particularly those crafted in pre-modern times—were integrated into the Chinese political, cultural, and economic systems. This requires a specific focus on the social status of craftsmen and how their work was perceived and valued within society.

3.5 Constructively Critical: A Confucian Philosophy of Technology

Currently, most scholarly works treat Chinese traditions as a monolithic whole, blending Daoism, Buddhism, and Confucianism, and sometimes even including Moism. However, we believe this approach is flawed. Confucianism, Daoism, and Buddhism, though often grouped together as ancient Chinese traditions, are quite distinct from one another and play very different roles in shaping Chinese society and thought.

Confucianism is generally seen as a practical, down-to-earth philosophy that has significantly shaped the mentality of the literati, especially in public service. In contrast, Daoism and Buddhism provide spiritual support, offering solace in the face of political or personal challenges. Given these differences, Confucianism is perhaps the most relevant tradition for understanding and developing a foundational narrative for technology in China. This doesn't mean that Confucianism necessarily fosters the most efficient technological culture; Moism and Daoism, with their focus on craftsmanship and alchemy, may have played a crucial role in that regard.

Confucianism has exerted influence across different dynasties in China, making it the tradition that has remained largely uninterrupted for thousands of years. Despite changes in rulers and power structures, Confucianism has endured, giving it a unique strength compared to other traditions when constructing a foundational narrative for the philosophy of technology. In addition, after a brief disruption during the Cultural Revolution, Confucianism is now experiencing a revival, especially with President Xi Jinping's emphasis on combining Marxism with China's specific conditions and integrating the outstanding aspects of Chinese traditions (Han, 2021). Given this resurgence, constructing a Confucian narrative in the philosophy of technology has become an urgent task.

We believe that a Confucian narrative of the philosophy of technology can effectively address Mitcham's question regarding the positivity of Chinese scholars toward technology. A Confucian perspective on technology may not be as straightforwardly positive as some modern views. Confucianism offers a nuanced understanding of technology that differs significantly from the modern perception of technology as merely an instrument for increased productivity. In Confucian tradition, there is a clear distinction between creating and making artifacts—two distinct concepts. The making of artifacts is the task of craftsmen, who hold a specific, though not highly esteemed, social position within the Confucian hierarchy. This hierarchy places government officials (士) at the top, followed by farmers(农), craftsmen(工), and then businessmen(商).

In contrast, the creation of the prototype of all artifacts is highly valued, attributed to the sage kings who, through observing the cosmos and understanding its laws, initiated the making of these artifacts (Lü, 2022). This creative act, which I refer to as "Confucian creative agency(圣王制器)," aligns with the natural order and cosmological laws, making these artifacts more than mere tools—they are seen as reflections of the cosmos itself.

The artifacts created by the Sage King function effectively in the real world because they embody cosmic laws, which, in some respects, resemble modern physical laws. However, it is important to note that in ancient Chinese tradition, there was no clear distinction between physical laws and existential laws. The cosmos was not viewed as a universe existing objectively apart from human beings. Instead, it was perceived as a holistic, sacred unity in which all beings reside. The cosmos did not have an inside or outside, nor external or internal aspects; it was an integrated whole where everything is interconnected.

Therefore, technological artifacts in the Confucian tradition are not merely tools; they are considered materializations of the Dao (or "Way"). This perspective imbues technology with a existential dimension, reflecting the deep interconnectedness between humanity, nature, and the cosmos. Therefore, our interaction with technological artifacts should not be reduced to a mere master-using-tools dynamic. Instead, these artifacts are viewed as possessing their own form of agency, which mediates our experiences and ultimately contributes to the development of our moral agency. In this view, artifacts are not passive objects but active participants in shaping how we engage with and understand the world, influencing our values and ethical considerations (Wang & Wong, 2025; Wang, 2021).

3.6 Conclusion

We hold that a Confucian philosophy of technology can offer a more balanced and nuanced perspective on the role of technology in human life, addressing Mitcham's concerns about the often uncritical embrace of technology. Confucianism provides a unique framework that neither blindly celebrates technological advancement nor outright condemns it. Instead, it advocates for a thoughtful engagement with technology, rooted in the belief that artifacts and tools are not merely objects of utility but are deeply interconnected with ethical living and societal harmony.

Unlike approaches that focus solely on the instrumental value of technology, Confucianism evaluates artifacts through a dual lens. On one hand, it acknowledges their practical functions in enhancing human life and facilitating progress. On the other hand, it views them as carriers of moral and cultural significance. In this sense, artifacts are not neutral entities; they are embedded within a web of human relationships and responsibilities. For Confucianism, the use of technology must align with broader moral goals, fostering virtues such as benevolence (仁), righteousness (义), and propriety (礼). This perspective implies that technology is not an end in itself but a means to cultivate and express virtuous moral agency.

For example, Confucianism might approach the design and use of technology by asking how it contributes to the well-being of the family, the community, and society at large. Does it promote harmony, responsibility, and the flourishing of human potential? Or does it disrupt these values? Such questions reflect the Confucian emphasis on relational ethics, where the value of technology is judged not just by its functionality but also by its impact on human relationships and moral development.

Given the different focus, we will not present a fully developed theory of Confucian philosophy of technology in this chapter. However, this sneak peek serves as a starting point for further study.

References

Chen, C. (1982). The difference and unity of science and technology. *Guangming Daily*, January 1 (陈昌曙.科学与技术的差异与统一 光明日报.1982-1-1).

Han, Q. (2021). Comprehensively and deeply understanding the core significance and essence of the 'two integrations'. *Marxist Studies, 10*, 93–105. (韩庆祥. 全面深入理解 "两个结合" 的核心要义和思想精髓. 马克思主义研究, 2021, 10: 93–105.).

He, Z., Zhao, H., & Guo, H. (1978). A critique of the 'theory of science and technology as part of the superstructure' advocated by the gang of four. *Philosophical Research, 04*, 13–22. (何祚庥,赵红州,郭汉英.批判"四人帮"的"科学技术上层建筑论". 哲学研究, 1978, (04): 13–22.).

Hui, Y. (2019). *The question concerning technology in China: An essay in cosmotechnics*. MIT Press.

Ihde, D. (2008). *Expanding hermeneutics: Technology in a postphenomenological perspective* (L. Han, Trans.). Peking University Press. (唐·伊德 著, 韩连庆译. (2008). 让事物说话: 后现象学视野中的技术科学. 北京大学出版社.)

Ihde, D. (2012). *Technology and the lifeworld: From garden to Earth* (L. Han, Trans.) Peking University Press. (唐·伊德 著, & 韩连庆 译. (2012). 技术与生活世界:从伊甸园到尘世. 北京大学出版).

Illich, I. (2017) *Deschooling society*. 伊万·伊利奇, (2017). 去学校化社会 (汉英双语版) (吴康宁, 译). 中国轻工业出版社.

Jiang, Y. (2023). From philosophy of technology to philosophy of engineering: A paradigm shift in philosophical research. *Zhejiang Academic Journal, 5*, 65–72. (怡. (2023). 从技术哲学到工程哲学:一种哲学研究范式的转型. 浙江学刊(5), 65–72.).

Li, S. (2009). *Restating tradition: A holistic comparative study of the philosophy of technology*. China Social Sciences Press. (李三虎.重申传统：一种整体论的比较技术哲学研究[M].中国社会科学出版社,2009.).

Liu, Z. (1992). The new development of Marxist views on science and technology: On the principle that 'science and technology are the primary productive forces'. *Studies in Dialectics of Nature, 01*, 3–9. (刘则渊.马克思主义科学技术观的新发展——关于"科学技术是第一生产力"的原理.自然辩证法研究, 1992, (01): 3–9.).

Lü, M. (2022). Ritual and music in craftsmanship: Examining Confucian ideas of production through the narrative of the three sovereigns. *History of Chinese Philosophy, 1*, 19–24. (吕明烜.礼乐化制作:从三皇叙事看儒家的制作观念.中国哲学史, 2022(1):19–24.).

Serezoe, T., & Wu, W. (1978). The philosophy of technology in the Americas. *Philosophical Translations, 04*, 71–74. (T.赛雷佐埃 ,吴伟 .美洲的技术哲学. 哲学译丛,1978, (04): 71–74.).

Wang, Q. (2005). From skill to Dao: The concept of philosophy of technology in Chinese tradition. *Philosophical Research, 12*, 84–89. (王前."由技至道"——中国传统的技术哲学理念.哲学研究, 2005, (12): 84–89.).

Wang, X. (2021). Confucian ritual technicity and philosophy of technology. In *Harmonious technology* (pp. 10–28). Routledge.

Wang, X., & Wong, P. (2021). *Harmonious technology: A Confucian ethics of technology*. Routledge.

Wang, X., & Wong, P. H. (2025). Images, moral feelings, and rites: Engaging Confucianism with philosophy of technology. *Dao, 24*, 1–20.

Wu, G. (2016). *What is science?* Guangdong People's Publishing House. (吴国盛. (2016). 什么是科学. 广东人民出版社.).

Yang, M. (1991). The formation and characteristics of Deng Xiaoping's thought that 'science and technology are the primary productive forces'. *Social Sciences, 07*, 6–10. (杨明刚.邓小平"科学技术是第一生产力"思想的形成和特色.社会科学, 1991, (07): 6–10.).

Zao, C. (1978). Conference on philosophical issues in technical sciences held in East Germany. *Philosophical Translations, 01*, 67–68. (早晨.东德举行技术科学的哲学问题讨论会.哲学译丛,1978,(01): 67–68).

Zhang, L. (1994). Reflecting on the meaning of technology in the 'lifeworld': A preliminary exploration of Don Ihde's philosophy of technology. *Studies in Dialectics of Nature, 07*, 26–30. (张来举.在"生活世界"里反思技术的意义——伊德技术哲学初探.自然辩证法研究， 1994, (07): 26–30).

Zhao, T. (2018). The 'immediate concerns' and 'long-term worries' of the artificial intelligence 'revolution': An analysis from ethics and ontology. *The Development of Philosophy in China, 4*, 8. (赵汀阳. (2018). 人工智能"革命"的"近忧"和"远虑"——一种伦理学和存在论的分析. 哲学动态(4), 8).

Part III
Chinese and Western Comparative Philosophies of Technology

Chapter 4
Conviviality and Confucian Oneness: A Case for Mutual Learning

Pak-Hang Wong

Abstract Over the past few decades, Carl Mitcham has served as an important interlocutor for philosophy of technology between the English-speaking world and China. In reflecting on this engagement in his speech for the Lifetime Achievement Award of the Society for Philosophy and Technology, Mitcham has emphasized the "Western deficiencies" and the resulting impoverished technological imaginary and defended the value of Chinese philosophical resources in envisioning a better *global* technological imaginary. Mitcham is certainly right to highlight the need for non-Western perspectives, but his narrative of "how the West ought to *learn from* China" may overshadow the opportunities for mutual learning and enrichment. In this essay, I propose a more reciprocal approach to cross- and inter-cultural philosophy of technology: we should focus on—or, at least, begin with—specific philosophical and normative ideas in different cultural traditions and examine how they might complement one another. By starting with specific philosophical and normative ideas, I argue that we can cultivate a more nuanced and productive dialogue among different traditions. To illustrate this approach, I put into conversation two of Mitcham's inspirations: Ivan Illich and Confucianism. More specifically, I examine how Illich's conviviality and Confucian idea of oneness can inform and enrich each other, thereby offering a richer vision for the technological imaginaries they inspire.

Keywords Conviviality · Oneness · Intercultural philosophy of technology · Ethics of interconnectedness · Neo-Confucian philosophy · Ivan Illich

---

P.-H. Wong (✉)
Academy of Chinese, History, Religion and Philosophy, Faculty of Arts and Social Sciences, Hong Kong Baptist University, Kowloon, Hong Kong
e-mail: pakhangwong@hkbu.edu.hk

L. Checketts (ed.), *What Went Wrong with the West: Carl Mitcham's Critique of Western Philosophy of Technology*, Philosophy of Engineering and Technology 52, https://doi.org/10.1007/978-3-032-22214-5_4

4.1 Introduction

Over the last few decades, Carl Mitcham has contributed significantly to the philosophical exchange between the English-speaking world and China by serving as an interlocutor of philosophy of technology in the English-speaking world for Chinese scholars and by co-authoring with Chinese scholars and promoting the relevance and significance of Chinese philosophy for philosophical and ethical analysis of technology (see, e.g., Mitcham et al., 2018; Zhu & Mitcham, 2020; Yan et al., 2022; Wang & Mitcham, 2024). In a keynote speech for the Lifetime Achievement Award of the Society for Philosophy and Technology in 2021, Mitcham highlighted "four Western deficiencies," namely the lack of agrarian life, split world ontology, negative gods, and historical discontinuity. Mitcham viewed Chinese cultural beliefs to be remedies of these deficiencies (Mitcham, 2026), and argued for the importance of Chinese philosophy in offering an alternative technological imaginary—or, indeed, a counter-imaginary—to the one that has dominated "the West." While insightful, I contend that Mitcham's narrative has overgeneralized the West, even if doing so is only for rhetorical purpose, and the overgeneralization risks burying valuable philosophical resources for creating a better technological imaginary *for the global world*.

Instead of framing the narrative as one about how the West ought to *learn from* China, I call for a more reciprocal account that emphasizes opportunities for mutual learning and enrichment. This requires moving beyond (over)generalized talk of "the West" or "China," which may obscure the richness and diversity within each tradition. Indeed, this is a methodological point I want to make approaching cross- and inter-cultural philosophy of technology: we should focus on—or, at least, begin with—specific philosophical and normative ideas in different cultural traditions and examine how they might complement one another. By starting with particular philosophical and normative ideas, we can cultivate a more nuanced and productive dialogue among different traditions, or so I shall attempt to demonstrate. In this spirit, I shall put into conversation two of Mitcham's inspirations: Ivan Illich and Confucianism. More specifically, I focus on Illich's idea of conviviality, which Mitcham has interpreted and applied in various contexts (Mitcham, 1991, 1996, 2009, 2018) and the Confucian idea of *oneness*, i.e., a central notion in Confucian philosophy (see Ivanhoe, 2015, 2017; Ivanhoe et al., 2018; Wong, 2021; Cocks, 2024), and illustrate how they can inform one another, thereby offering a richer vision for the technological imaginaries they inspire.

4.2 Carl Mitcham on Ivan Illich's Concept of Conviviality

Mitcham has been a major proponent of Ivan Illich, a social critic who wrote extensively on the impacts on human relationships and well-being from the inherent characteristics of modern technologies and institutions (Mitcham, 1991, 2003).

Particularly, Illich has developed the notion of conviviality in *Tools for Conviviality* (1973) in response to the phenomenon of counterproductivity of modern technologies and institutions, where technologies and institutions grow beyond their inherent limits and no longer serve basic human needs they are initially intended for but instead increase "regimentation, dependence, exploitation, and impotence" (1973: 20; see Voinea, 2018; Mitcham, 2003). Illich illustrates the counterproductivity of modern technologies with examples such as medicine and iatrogenic illness and cars and traffic jams, amongst others. As Mitcham observes, Illich's argument primarily focuses on the characteristics—or, the inner structures—of technologies, and thus offers another way to critically evaluate *technology itself* as opposed to technology-driven *human actions* (Mitcham, 1991: 39–40).[1] He contends that people ought to have "the freedom to [create technologies] among which they can live, to give shape to them according to their own taste and to put them to use in caring for and about others" (Illich, 1973: 11). So construed, Illich's main concern is not about specific uses of technologies or the intention behind those uses, but about the conditions for personal freedom and caring opened up or closed down by technologies. In other words, it is about the ethics and politics of design and integration of technology to foster personal freedom and care.

Central to Illich's argument is the normative concept of conviviality, which refers to "autonomous and creative intercourse among persons, and the intercourse of persons with their environment; and this is in contrast with the conditioned response of persons to demands made upon them by others, and by a man-made environment" (Illich, 1973: 11). So defined, conviviality emphasizes autonomy, creativity, and relationality with one another and with our environment. More specifically, Illich looks to identify the kind of technology that supports conviviality, thereby distinguishing between convivial tools that "give each person who use them the greatest opportunity to enrich the environment with the fruits of his or her vision", and non-convivial tools that "deny this possibility to those who use them and they allow their designers to determine the meaning and expectations of others" (Illich, 1973: 21). Convivial tools, therefore, are technologies that allow users to use them without the supervision of external authority and realize what is meaningful to the users themselves. As Mitcham points out, it is important not to mistake Illich as advocating freeing oneself *from others* with technologies in convivial tools given his emphasis on people's interdependence with one another and with the environment (Mitcham, 2009: 306). For Illich, it is certainly the case that technologies ought not to engender hierarchical control over their users, but his emphasis on conviviality as being about

[1] Illich's concern in *Tools of Conviviality* is "tool", but he use the term "broadly enough to include not only simple hardware such as drills, pots, syringes, brooms, building elements, or motors, and not just large machines like cars or power stations; I also include among tools productive institutions such as factories that produce tangible commodities like corn flakes or electric current, and productive systems for intangible commodities such as those which produce 'education,' 'health,' 'knowledge,' or 'decisions' (1973: 20). In other words, Illich's analysis does not distinguish between modern technologies and social institutions. In this paper, I shall use the term "technology" in lieu of Illich's "tool" to avoid lack of clarity.

"the intercourse among persons" and "the intercourse of persons with their environment" should remind us not to understand autonomy as freedom from others but as freedom *to engage with* and *for others* through technologies.

For technologies to be convivial, they must be

> easily used, by anybody, as often or as seldom as desired, for the accomplishment of a purpose chosen by the user. The use of such tools by one person does not restrain another from using them equally. They do not require previous certification of the user. Their existence does not impose any obligation to use them. They allow the user to express his meaning in action (Illich, 1973: 22).

Hence, convivial technologies should *not* require excessive, specialized, or professionalized training, as the training limits users' autonomy by reducing the availability of other options to the users or by restricting their access to the technologies. Convivial technologies also should *not* impose obligations on users insofar as the imposition of any obligation entails both a reduction of autonomy and an enactment of external control over the users. Alternatively, technologies that are non-convivial—or even *anti*-convivial—are those characterized by hierarchical and centralized control, socio-technological opacity, and dependency of users.

Illich's concept of conviviality and the related notions such as convivial tools, therefore, offer several normative criteria for critical reflection of technologies, that is—whether a technology under critical scrutiny is easily usable, whether it imposes any obligation on its users—be the obligations physical, cognitive, social, or technical—and, if so, what is the nature of the obligation imposed. Does it limit users' autonomy or put them under any external authority? Alternatively, we should reflect on whether the technology engender hierarchical, centralized control, perhaps, through radical monopolies such as the introduction of cars monopolizing urban transit by pushing out other forms of transportation, or the school system monopolizing education by rejecting and/or devaluing self-education, apprenticeships, and other forms of non-institutionalized means of learning (see Mitcham, 1991: 27–28). Another normative concern is whether technologies are opaque, leaving their users unaware of how they function, thereby promoting deference to external authorities who know 'better' about how they should engage with the technologies; or, indeed, whether the technologies encourage their users to relinquish their natural abilities in exchange for 'better' performance through their reliance on technologies.

Extending Illich's insights, Mitcham (2009) examines the free and open-source software (FOSS) movement through the concept of conviviality and the related notion of convivial tools. Mitcham notes that, from the purview of conviviality, the FOSS movement should not merely occupy itself with questions about the accessibility of software or the availability of the technical source code, but should concern with the characteristics or the inner structures of the software that allows users to have meaningful engagement with and through the software, i.e., convivial software. He proposes three characteristics for software that can be viewed as convivial, namely (1) stability, which enables users to maintain long-term relations with the software tools by relieving them from the need for continuous upgrades, which limit users' control over their ways of working and force changes on them; (2)

transparency, which allows users to understand the functioning of software tools, manipulate the tools to serve their ends, and address the errors and malfunctions by themselves, thereby reducing unnecessary deference to experts; and, (3) simplifiability, which opens up the possibilities for users to shape and reshape their ways of working with and through their software tools, thereby allowing for genuinely meaningful user experience (Mitcham, 2009: 307).

It is worth noting that Mitcham's criteria apply beyond FOSS and more generally to digital technologies that run on software. In this respect, digital technologies, including digital apps and platforms, can also be evaluated via the three criteria of stability, transparency, and simplifiability. Indeed, as the role of digital technologies, such as algorithmic systems and AI-driven applications, become increasingly prominent in people's everyday life, shaping and structuring their experience. Mitcham's criteria offer another way to critically evaluate today's digital technologies. Many of these apps and platforms require constant updates—often implemented with only minimal awareness from users, such as through auto-update—and they are also becoming increasingly opaque in how they function and why they fail when they do, particularly in the case of AI. Much like Illich's critique of non-convivial technologies, today's digital technologies promote centralized control, opaqueness, and user-dependency. For example, in the case of AI, they are often advertised as indispensable and inevitable, reinforcing the image that people *must* adopt AI or risk obsolescence. Hence, these digital technologies would most certainly be critiqued by Illich or Mitcham as *anti-convivial.*

Finally, it is crucial to not understate the political dimension of conviviality, as articulated by both Illich and Mitcham. For Illich, conviviality necessitates what he calls a "[a] new politics of tools would aim principally to exclude the design of artifacts [...] that are obstacles to the exercise of [...] personal freedom. Such politics would limit the scope of tools as demanded by the protection of three values: survival, justice, and self-defined work" (Illich, 1973: 13). Similarly, Mitcham's call to integrate the criteria into software tools and digital technologies must also be viewed as a political project—one that involves not only designing and developing tools with stability, transparency, and simplifiability, but also must examine and challenge the lack of the characteristics of conviviality in the existing digital technologies (see, e.g., Mitcham, 1991: 37).

4.3 Confucian Oneness and the Ethics of Interconnectedness

Confucianism shares with Illich's (and Mitcham's) notion of conviviality a fundamental ethical vision, namely an emphasis on relationality, interdependence, and mutual flourishing. While Illich and Mitcham apply the concept of conviviality to critique and reflect on non-convivial technologies' propensity to diminish the freedom to engage *with* and *for* others through technologies, Confucianism, particularly the Confucian notion of oneness, challenges the (hyper-)individualistic view of the self in contemporary Western ethics and political philosophy, and advocates an expanded view of the self that encompasses other people, non-human creatures, and nature (Ivanhoe, 2017). As Illich's (and Mitcham's) notion of conviviality calls for a new politics of design and integration of technologies that focus on their characteristics and inner structures, Confucian oneness provides an alternative ontological foundation to philosophical and ethical reflection of technology based on an ethics of interconnectedness. In this section, I shall introduce the Confucian notion of oneness and highlight some of its implications to philosophical and ethical analysis of technology before turning to an exchange between conviviality and Confucian oneness in the next section.[2]

Oneness of Heaven and Humanity (*tianren heyi* 天人合一) is the normative ideal in Confucian philosophy, which specifies the ideal relationships between human and nature as well as between humans themselves. This Confucian notion of oneness is best exemplified by the Neo-Confucian thinker Zhang Zai's *Western Inscription*:

> Heaven is my father and Earth is my mother, and even such a small creature as I find an intimate place in their midst. Therefore *that which fills the universe I regard as my body and that which directs the universe I consider as my nature. All people are my brothers and sisters, and all things are my companions* (Chan, 1963: 497, author's emphasis).

The notion of one body (and, relatedly, one family) is central to Zhang Zai's philosophy, and to Neo-Confucian philosophy more broadly. Here, the description of the relationship between human and nature as "one body" is not intended to be metaphorical but metaphysical. Zhang Zai and other Neo-Confucian thinkers hold the metaphysical claim that the self is in an important aspect *identical* to all things in the world, and they conceive all human beings as sharing the same substance (or a common pattern) and being one with and through the universe (see Angle & Tiwald, 2017). This metaphysical claim grounds a deep and meaningful connection among people, non-human creatures, and all things in the world, and thus people's flourishing is inherently interconnected, i.e., that one's flourishing inherently depends on the flourishing of others, non-human creatures, and all things in the world. So construed, one cannot meaningfully distinguish between one's own flourishing and the flourishing of others. In this respect, it is people's interconnectedness with the universe and each other that we ought to tend to ourselves as tending to our universe and other beings. Confucian oneness requires people to care for the universe and other beings, because failing to do so will be detrimental to the flourishing of the universe (if not downright harmful) and others, which undermines the basis of their own flourishing. Similarly, people's failure to care for themselves, thereby harming themselves, will harm the universe and other beings insofar as they are one body. In other words, Confucian oneness foregrounds the normative significance of

[2] Parts of this section are derived from an earlier paper, "Artificial Intelligence, Personal Decisions, Consent, and the Confucian Idea of Oneness" (Wong, 2021). In Wong (2021), the aim is to introduce the Confucian notion of oneness as an *additional* ground for reconceptualizing moral responsibility as a *collective* notion. In this section, my aim is to propose an ethics of interconnectedness based on Confucian oneness.

interconnectedness and requires us to reconsider our moral responsibility to nature and each other.

P. J. Ivanhoe summarizes this Neo-Confucian philosophical view with the oneness hypothesis, which "is a view about the nature of the world; its primary moral aspect concerns the nature of the relationship between the self and the other people, creatures, and things of the world; its core assertion is the claim that we—and in particular our personal welfare or happiness—are inextricably intertwined with other people, creatures, and things" (Ivanhoe, 2017: 1; also, see Ivanhoe, 2015; Ivanhoe et al., 2018). Ivanhoe (2015) argues that the Confucian oneness leads to an expanded view of the self and an enlarged sense of responsibility. According to Neo-Confucianism, as we come to see that there is no distinction between the self and others, we recover a deep connection between us and other people, non-human creatures, and all things in the world. This deep connection allows us to understand and feel for them, and the deep connection also motivates and obliges us to care for them. In an important sense, one's failure to care for others and the world does not only fail *them* morally, but it morally fails oneself—it is a form of self-harm and self-degradation. This expanded view of the self, in effect, calls for us to actively promote the flourishing of other people, non-human creatures, and the nature. Hence, *The Doctrine of the Mean* states,

> Only those who (are) absolutely sincere can fully develop their nature. If they can fully develop their nature, they can then fully develop the nature of others. If they can fully develop the nature of others, they can fully develop the nature of things. If they can fully develop the nature of things, they can assist in the transforming and nourishing process of Heaven and Earth, they can thus form a trinity with Heaven and Earth (The Doctrine of the Mean 22, in Chan, 1963: 107–08).

This Neo-Confucian ethics of interconnectedness, therefore, challenges philosophical and ethical frameworks that draw a sharp distinction between self-regarding and other-regarding actions. We can even abandon the difference between the moral significance of self-regarding and other regarding actions and consider them to be on par morally. Accordingly, the Neo-Confucian ethics of interconnectedness posits that *all* ethical actions must account for their broader impacts on the interconnected web of relations.

The Confucian notion of oneness presents the thesis that we *are* and *should* be at one with others and the world, but why, in reality, people often do *not* care for others? In other words, what prevents people from achieving Confucian oneness? Neo-Confucian thinkers such as Zhu Xi and Wang Yangming believe that self-centeredness is the primary obstacle to achieving oneness (Tien, 2012; Ivanhoe, 2017, Ivanhoe et al., 2018; Tiwald, 2018). In his interpretation of Wang Yangming's account of how to achieve oneness, Ivanhoe distinguishes between selfishness and self-centeredness, and defends the view that the Neo-Confucian thinker takes the latter to be the main obstacle to achieving oneness (Ivanhoe, 2017: 66–73; also see Tien, 2012). Unlike selfishness, which is interpersonal in nature—being selfish is to prioritize one's own concerns and interests over and against the concerns and interests of others—self-centeredness denotes a specific mode of being towards others and the world that a person takes oneself to be the center. In doing so, one "[takes] the world excessively and exclusively from one's own point of view and to view oneself as distinctly separate from other persons and things" (Tien, 2012: 53). Since self-centeredness prioritizes one's own perspective over any other perspectives, and as prioritizing one's view and centered on it draws a sharp distinction between the self and others, self-centeredness makes it difficult for people to recognize and experience the interconnection with others and the world, thereby also making it difficult for them to achieve oneness. Accordingly, self-centeredness is *the* obstacle to achieving oneness, and addressing self-centeredness is one of the key issues among Neo-Confucian thinkers.

I have thus far identified two components in the Neo-Confucian ethics of interconnectedness, namely an enlarged sense of responsibility that rejects the distinction between self-regarding and other-regarding actions and the moral significance to avoid—or eliminate—self-centeredness. Both the enlarged sense of responsibility and the concern for self-centeredness can inform our normative analysis of technology.[3] For instance, Confucian oneness, for its focus on interconnectedness amongst people and between people and the world, requires technology to provide the conditions for mutual flourishing and collective well-being in lieu of *personal flourishing* or *individualistic well-being*. The Neo-Confucian ethics of interconnectedness therefore calls for 'pro-oneness technologies' that enable proximal relations, openness, deep and meaningful embeddedness, and solidarity. At the same time, since the main obstacle to achieving Confucian oneness is self-centeredness, the Neo-Confucian ethics of interconnectedness will critique technology—or, technological design—that is distancing, (self-)enclosing, decontextualizing, and isolating, thereby encouraging self-centeredness.

Building on Mark Coeckelbergh's (2015a, 2015b, 2015c) insights regarding how (digital) technologies can create physical, experiential, social, and moral distance that diminishes our capacities for thinking and feeling for distanced others, the Neo-Confucian ethics of interconnectedness further emphasizes how such distancing fundamentally severs our deep connection with others and the world essential for overcoming self-centeredness. The capacities of thinking and feeling with others and the world are key to dissolving the perceived separation between the self and others as well as self and world. Currently, there are various (digital) technologies like AI personalization that create epistemic and affective bubbles, thereby imposing boundaries on individuals and restricting their cognitive and affective reach—a reach essential to the elimination of self-centeredness. Pro-oneness technologies should function differently: they should be designed to nurture the cognitive and affective reach, promoting engagement with diverse perspectives and experiences. For example, the art project *In the Eyes of the Animals* (intheeyesoftheanimal.com) created by Marshmallow Laser Feast (MLF) uses virtual reality technology to help the audience to expand their cognitive and affective reach by allowing them to

[3] Elsewhere, I have discussed how Confucian oneness supports a reconceptualization of moral responsibility of the users of AI-driven algorithmic systems (see, Wong, 2021).

experience the sensory perspectives of a midge, a dragonfly, a frog, and an owl (Boyacioglu & Ersin, 2017).

Technologies that detach individuals from their social, cultural, and historical traditions likewise undermine a sense of belonging crucial for an expanded self and resistance to self-centeredness. In this respect, pro-oneness technologies should be direct us away from the possibility of detachment by (digital) technologies and counter this by foregrounding the importance of embeddedness and social, cultural, and historical continuity in technology design and implementation. Finally, when technologies isolate people from others and the world, they also encourage people to take themselves as the center of the world by removing other possible perspectives available. For instance, addictive and manipulative technologies monopolize users' attention and leave them no mental capacities to relate to anything else other than those offered by the technologies (cf. Bhargava & Velasquez, 2021; Schuster & Lazar, 2025). In this respect, pro-oneness technologies ought to be designed to minimize the risks of technology-induced addiction and manipulation.

The Neo-Confucian ethics of interconnectedness offers a different perspective to normative analysis of technology by drawing our attention to technology's implications for oneness. This Neo-Confucian perspective foregrounds the normative significance of technologies being designed to enable proximal relations, openness, deep and meaningful embeddedness, and solidarity. Conversely, technologies that act as obstacles to oneness—by creating distance, enclosing users, decontextualizing experience, or isolating individuals—are seen as hindering the realization of Confucian interconnectedness. In short, when applying the Neo-Confucian ethics of interconnectedness to normative analysis of technology, we can assess technologies through the lens of pro-oneness technologies and self-centered technologies (or, anti-oneness technologies).

4.4 A Case for Mutual Learning Between Conviviality and Confucian Oneness

Illich's (and Mitcham's) notion of conviviality and the Neo-Confucian-inspired normative analysis of technology I have proposed in the previous section share a fundamental concern for the role of technology in structuring and shaping human relationships: Illich's critique of non- and anti-convivial tools aims to show how technologies, through their characteristics and inner structures, can constrain people's autonomy and limit meaningful participation in one's own community, whereas the Neo-Confucian concern over self-centeredness allows us to critique technologies for their being detrimental in cultivating dispositions and capabilities to recognize and honor (Confucian) oneness. By putting them into conversation, I contend that we can advance a more holistic normative analysis of technology. In this section, I elaborate on *four* ways they can complement one another. More specifically, I shall argue that (a) Confucian oneness supplements the notion of conviviality a robust account of moral responsibility, (b) conviviality counterbalances the paternalistic tendency inherent in the Neo-Confucian ethics of interconnectedness; (c) the notion of self-centeredness helps account for counterproductivity and non-convivial technologies; and (d) conviviality offers possible ways to politically structure and institutionalize Confucian oneness in normative analysis of technology, and thus politically operationalizes the Neo-Confucian ethics of interconnectedness.

The notion of conviviality focuses on critiquing the inherent characteristics and inner structures of non-convivial technologies, and identifies the convivial features such as being non-hierarchical, decentralized, accessible, transparent, or democratic that we should design and integrate in our technologies, but it lacks a robust account of moral responsibility to account for *why we ought to design technologies in convivial ways and/or enact self-limiting to avoid counterproductivity* beyond Illich's anthropology of humans as cohabitants, which assumes both freedom to engage with and care for others and meaningful participation are inherently good (see, e.g., Mitcham, 1991). In this respect, Confucian oneness can fill the conceptual gap by offering a robust account of moral responsibility based on a strong ontological foundation, i.e., people's interconnectedness at the fundamental, metaphysical level, and answers why we ought to design for convivial technologies: convivial technologies are technologies that enable individuals to engage with and care for each other, which are essential to recognize and honour our interconnectedness. Hence, as Confucian oneness emphasizes the interconnectedness among people and between people and the world, and thus one's flourishing depends on the flourishing of the embedded web of relationships, this relational ontology justifies the normative significance of designing technologies in convivial ways because of its importance for mutual flourishing. Similarly, conviviality's call for self-limitation can also be justified by the Neo-Confucian ethics of interconnectedness: counterproductivity leads to regimentation, dependence, exploitation, and impotence, which does not only limit personal freedom and human potential but also disrupts the conditions for cultivating dispositions and capabilities for achieving oneness.

Although the Confucian notion of oneness does supply a robust account of moral responsibility and a strong ontological foundation to conviviality, one should be wary of the paternalistic tendency inherent in its emphasis on interconnectedness to others and the world, in which their concerns and interests can take precedence over one's own. Accordingly, individuals' agency—their autonomy and opportunities for meaningful participation—risk being subordinated to projects of mutual flourishing and collective well-being. Indeed, the idea of pro-oneness technologies proposed in the previous section, which aims to foster proximal relations, openness, deep and meaningful embeddedness, and solidarity is *in principle* compatible with non-convivial and anti-convivial technologies that impose centralized control and user dependency, particularly when they are done in the name of oneness. Here, conviviality counterbalances the paternalistic tendency by insisting on the preservation of individual agency within a relational ontology and so must not be eliminated entirely. In other words, while the Neo-Confucian ethics of interconnectedness rightly emphasizes one's broader responsibility towards others and the world, conviviality reminds us that the enlarged sense of responsibility ought not to result in centralized control and dependency that diminish individuals' agency, thereby calling for safeguards against excessive prioritization of the obligations to mutual flourishing and collective well-being.[4]

Self-centeredness, which is identified by Neo-Confucian thinkers as the main obstacle to oneness, may offer an account for the rise of counterproductivity and the prominence of non-convivial technologies. Neo-Confucian thinkers suggest that self-centeredness prevents people from considering perspectives beyond themselves and thus blinds them from the broader social and ethical impacts of their actions. When people approach technology in a self-centered manner, they will not concern themselves with its outcome at the collective level, so long as the technology is useful and/or beneficial to themselves even if only in short-term, which, in turn, helps explain the unchecked growth of technology resulting in counterproductivity and the adoption of non-convivial technologies, especially when the non-convivial technologies enable seemingly better performance and results. Relatedly, self-centered technologies, as I have outlined in the previous section, dispose their users to be (more) self-centered, thereby making them more susceptible to counterproductivity and non-convivial technologies. From the perspective of the Neo-Confucian ethics of interconnectedness, counterproductivity and non-convivial technologies thrive because technologies are designed and adopted in ways that reinforce self-centeredness. Similarly, the idea of pro-oneness technologies, which cultivate users' dispositions and capabilities for achieving oneness should help respond to counterproductivity and non-convivial technologies.

Finally, the political dimension of conviviality deserves further attention, as it highlights the role of politics in shaping technologies by guiding the features to be designed and integrated in technologies. Illich's critiques of professionalization and radical monopolization are examples of how political structures contribute to the rise and prominence of non-convivial technologies. While I have proposed the idea of self-centered technologies to describe the features we should exclude from the design and integration of technologies, and the idea of pro-oneness technologies for features we should designed and integrated in technologies, the Neo-Confucian-inspired normative analysis of technology lacks explicit mechanisms to explore the political questions such as professionalization, radical monopoly, and the participation in the technology design and integration processes. By integrating conviviality's modes of political engagement in the form of participatory and decentralized governance, the Confucian notion of oneness can be *politically* operationalized and so evolves beyond merely an ethical framework into a framework that shapes *governance* of technology. Conviviality, therefore, provides practical mechanisms for implementing the ethical vision of oneness through politics.

---

[4] It is, however, worth to point out that the discussion in Confucian political philosophy has explored similar questions related to (virtuous) paternalism and a number of responses have been proposed. For example, by means of "self-restriction" (see Wong, 2013: 361–362). It is beyond the scope of this paper to discuss these questions, but they can be leveraged to address the problem of (virtuous) paternalism.

4.5 Conclusion

Through a discussion of Illich's (and Mitcham's) notion of conviviality and the Confucian notion of oneness, I have highlighted the potential for mutual learning between Western and Chinese philosophy for normative analysis of technology and elaborated on a specific approach to inter- and cross-cultural philosophy of technology. Mitcham is certainly right to advocate for greater recognition of Chinese philosophy's role in reimagining (global) technological imaginaries, but his framing of the West *learning from* China risks overgeneralization and neglects valuable resources from within the West. This is why, I contend, a reciprocal approach to inter- and cross-cultural philosophy of technology that emphasizes *exchange* between philosophical and cultural traditions is more desirable.

For instance, conviviality emphasizes the need for technologies that preserve personal autonomy and meaningful participation, while Confucian oneness extends this discussion by emphasizing our moral responsibility toward others and the world. Moreover, conviviality provides a crucial counterbalance to the potential paternalism inherent in the Neo-Confucian ethics of interconnectedness, ensuring the projects of mutual flourishing do not override individual agency. Conversely, the Neo-Confucian critique of self-centeredness offers a plausible explanation for the prominence of non-convivial technologies, reinforcing the need to design and integrate technologies that foster interdependence rather than isolation.

My focus in this paper has been to explore the complementarity between conviviality and Confucian oneness, and thus I have left open questions regarding their dissimilarities and incompatibilities. I have also not explored in detail how a holistic normative analysis of technology based on the insights of conviviality and Confucian oneness works in practice. They are, without doubt, important questions that deserve further investigation. However, the aim of this paper is not to offer a comprehensive approach to normative analysis of technology based on conviviality and Confucian oneness, but to demonstrate how an analysis of normative concepts from different philosophical and cultural traditions can mutually enrich one another, thereby contributing to a richer, more pluralistic and globally inclusive philosophy of technology, which, I contend, is a project Mitcham will endorse.

References

Angle, S. C., & Tiwald, J. (2017). *Neo-Confucianism: A philosophical introduction*. Polity.

Bhargava, V. R., & Velasquez, M. (2021). Ethics of the attention economy: The problem of social media addiction. *Business Ethics Quarterly, 31*, 321–359.

Boyacioglu, B., & Ersin, E. H. (2017). *Interview with Marshmallow laserfeast*. MIT Open Documentary Lab. https://docubase.mit.edu/lab/interviews/interview-with-marshmallow-laser-feast/

Chan, W.-T. (Trans.) (1963). *A source book in Chinese philosophy*. Princeton University Press.

Cocks, S. (2024). A Neo-Confucian architectural ethic. *Dao, 23*, 449–470.

Coeckelbergh, M. (2015a). *Money machines: Electronic financial technologies, distancing, and responsibility in global finance*. Ashgate.

Coeckelbergh, M. (2015b). *Environmental skill: Motivation, knowledge, and the possibility of a non-romantic environmental ethics*. Routledge.

Coeckelbergh, M. (2015c). The tragedy of the master: Automation, vulnerability, and distance. *Ethics and Information Technology, 17*, 219–229.

Illich, I. (1973). *Tools for conviviality*. Harper & Row.

Ivanhoe, P. J. (2015). Senses and values of oneness. In B. Bruya (Ed.), *The philosophical challenge from China* (pp. 231–251). MIT Press.

Ivanhoe, P. J. (2017). *Oneness: East Asian conceptions of virtue, happiness, and how we are all connected*. Oxford University Press.

Ivanhoe, P. J., Flanagan, O., Harrison, V. S., Sarkissian, H., & Schwitzgebel, E. (2018). *The oneness hypothesis: Beyond the boundary of self*. Columbia University Press.

Mitcham, C. (1991). Tools for conviviality: Argument, insight, influence. In P. T. Durbin (Ed.), *Europe, America, and technology: Philosophical perspectives* (pp. 17–56). Springer.

Mitcham, C. (1996). The argument of tools for conviviality and beyond. *Bulletin of Science, Technology & Society, 16*, 246–251.

Mitcham, C. (2003). In memoriam Ivan Illich: Critic of professionalized design. *Design Issues, 19*, 26–30.

Mitcham, C. (2009). Convivial software: An end-user perspective on free and open source software. *Ethics and Information Technology, 11*, 299–310.

Mitcham, C. (2018). Teaching with and thinking after Illich on tools. *The International Journal of Illich Studies, 6*(1).

Mitcham, C. (2026). Re-imagining Chinese technological imaginaries. In L. Checketts (Ed.), *What went wrong with the West: Carl Mitcham's critique of western philosophy of technology*. Springer.

Mitcham, C., Li, B., Newberry, B., & Zhang, B. (2018). *Philosophy of engineering, East and West*. Springer.

Schuster, N., & Lazar, S. (2025). Attention, moral skill, and algorithmic recommendation. *Philosophical Studies, 182*, 159–184.

Tien, D. W. (2012). Oneness and self-centeredness in the moral psychology of Wang Yangming. *Journal of Religious Ethics, 40*, 52–71.

Tiwald, J. (2018). Two notions of empathy and oneness. In P. J. Ivanhoe, O. Flanagan, V. S. Harrison, H. Sarkissian, & E. Schwitzgebel (Eds.), *The oneness hypothesis: Beyond the boundary of self* (pp. 371–387). Columbia University Press.

Voinea, C. (2018). Designing for conviviality. *Technology in Society, 52*, 70–78.

Wang, N., & Mitcham, C. (2024). What Western philosophers of technology might learn from Li Bocong's philosophy of engineering. *Science and Engineering Ethics, 30*, 32.

Wong, P.-H. (2013). The public and geoengineering decision-making: A view from Confucian political philosophy. *Techné: Research in Philosophy and Technology, 17*, 350–367.

Wong, P.-H. (2021). Artificial intelligence, personal decisions, consent, and the Confucian idea of oneness. In T. X. Wang & P.-H. Wong (Eds.), *Harmonious technology: A Confucian ethics of technology* (pp. 78–94). Routledge.

Yan, P., Wang, Q., & Mitcham, C. (2022). Taking responsible innovation to China: The Dalian Port development case. *Science and Engineering Ethics, 28*, 4.

Zhu, Q., & Mitcham, C. (2020). Liu Zeyuan's philosophy of engineering and technology: An introduction to his Marxist socioeconomic theory. *Technology in Society, 63*, 101351.

Chapter 5
Technology, Science, and Modernization: Progressive Tales and Modern Myths

Alfred Nordmann

Abstract There have been many proposals to move beyond the Needham-question and to treat Chinese technology on its own terms. In order to do so, it is necessary to abandon the notion that technological development participates in an overarching historical process which is sometimes referred to as *Technisierung* or "technification." While such a notion of progressive change and an ever more pervasive reach agrees with "modernization" and *Verwissenschaftlichung* or "scientification," it applies to technological development only to the extent that one considers science-driven technologies. However, this narrowing of perspective blinds us to the technical civilization of seventeenth century China, and blinds us also to technoscientific playfulness, to deeply entrenched technical traditions, and the varied creative expressions of *homo faber* in the twenty-first century. The challenge of separating out the notions of technification and scientification is exemplified by a discussion of Gernot Böhme's philosophy of technology. What went wrong with the philosophy of technology in the West was the Needham-question itself and with it the fixation on modern science at the expense of technical and technoscientific ways of knowing the world.

Keywords Technisierung · Gernot Böhme · Francis Bacon · Song Yingxing · China

There is a complementary German version of this essay. This chapter highlights the challenge of moving from the parochialism of the Needham question to a comprehensive political philosophy of technology. The German version is primarily dedicated to the discussion of Gernot Böhme's philosophy of technology which in the present version serves only to exemplify the magnitude of the "Needham predicament" (Nordmann, 2025).

A. Nordmann (✉)
Technical University of Darmstadt, Darmstadt, Germany
e-mail: nordmann@phil.tu-darmstadt.de

L. Checketts (ed.), *What Went Wrong with the West: Carl Mitcham's Critique of Western Philosophy of Technology*, Philosophy of Engineering and Technology 52, https://doi.org/10.1007/978-3-032-22214-5_5

5.1 Bewitched by the Needham Question

Thinking about science and technology, modernity and progress in China and the Western World—for the last century at least, this amounted to thinking about a deficit. Paradoxically, perhaps, it was associated with the name of a historian who wanted to dispel the myth of such a deficit. In the larger picture of what Joseph Needham was up to, the out-sized Needham question identified a small puzzle only. But in a certain way of thinking, this small puzzle provides the key to the biggest problems and questions of the modern world. This way of thinking is actually quite peculiar, a powerful prejudice, and though it was criticized and exposed now and then, it proved pervasive in East and West, among promoters and critics of Western civilization alike.

Consider the following three historical propositions on China and the West. Together they circumscribe what one might call the Needham predicament:

1. Joseph Needham went to great lengths to show that China has been leading the way in terms of technical inventiveness: "Gunpowder, the magnetic compass, and paper and printing, which Francis Bacon considered as the four most important inventions facilitating the West's transformation from the Dark Ages to the modern world, were invented in China" (Lin, 1995, p. 270).[1] These four, the "seven great inventions," and many more have been part of Chinese lore, every school-child knows them, and yet in the modernization narrative they do not count for much.
2. Needham appeared puzzled by the fact that at the time of Bacon and Galileo something happened only in the West, namely the emergence of scientific thinking: "Why did modern science, the mathematization of hypotheses about Nature, with all its implications for advanced technology, take its meteoric rise only in the West at the time of Galileo [but] had not developed in Chinese civilisation or Indian civilisation?" (Needham, 1969, p. 16, see 1954, p. 13). Interestingly enough, Needham does not speak about modern science in general terms. A biochemist himself, he refers to a highly specialized methodological practice, namely the employment of mathematized hypotheses about nature. This is a practice that responds to what one may call ontological anxiety—the fear that we cannot truly know reality but only the constructions of our own mind. This is not anxiety or fear in a personal or psychological sense. As Stanley Cavell showed with particular poignancy, to conquer ontological anxiety is the central

[1] Needham here adopts and restitutes Bacon's narrative. Bacon himself wonders about the obscure origin of these inventions. For him they signify the achievements that set the modern world apart from the scholastic debates of ancient and medieval science. He does not seem to care much where they came from, or whether one takes them as preparing or attending the emergence of modern science (Bacon, 2000, aph. 129, p. 100). According to Poor (2020), Needham helped China reclaim this narrative, though without the chapter on modern science.

concern of an epistemology that seeks to answer the skeptics (Cavell, 1979). The fact of modern science and the tradition of empiricism offers not a conclusive but a very good provisional answer to the skeptics, perhaps the best there is: As for well-made constructions of the mind, mathematics provides them and so we endeavor to recover bits and pieces of an unknown reality by casting out perfectly intelligible, mathematically constructed nets or hypotheses. These mathematized hypotheses can capture regularities to the extent that such regularities do exist. If this is all we can do—and from Bacon onward, this is what empiricists believed—it is actually not so little and takes considerable discipline to see it through.

3. Of course, this peculiar epistemic practice does not recommend itself where there is no such ontological anxiety in the first place—not in Chinese culture, for example, with its elaborate bureaucracy, well articulated technologies, and corresponding industries. But then, who would have thought that the disciplined scientific employment of mathematized hypotheses would be considered the key to the modern world? The very idea of scientific, technological, economic, cultural progress is inextricably linked to the accumulation of discovered regularities and attendant technological possibilities.

What by Needham's standards would be a small deficit—as formulated in the second of these propositions—turns out to be a huge deficit. The achievement recorded in the first proposition pales or proves insignificant in the light of modernization and the idea of progress of the third. Chinese culture awakened to this deficit in the mid-nineteenth century and defined itself aspirationally as trying to "catch up"—and has so well caught up by now that only recently it started asking again what a specifically Chinese way may look like and what the West has missed out on. Officially advocated since 2021, this renewed "historical confidence" might reveal that the specifically Chinese way is one from which it has never departed.[2] Accordingly, the detour through modern science with its mathematized hypotheses may have been overrated, given the realization in recent decades that the technoscientific expansion of knowledge in agriculture, nanotechnology, bioengineering, or computer science is not beholden to the empiricist methodologies of modern Western science (Carrier, 2011).

[2] The "four confidences" of President Xi's Chinese Socialism were expanded in November 2021 to include "historical confidence" in the *Resolution of the Central Committee of the Chinese Communist Party on the Major Achievements and Historical Experiences of the Party's Hundred-Year Struggle* (中共中央关于党的百年奋斗重大成就和历史经验的决议).—Without applying it to the Needham question explicitly, the programmatic implications of "cultural and historical confidence" are clearly indicated in a 2023 paper on "Understanding the Historical Significance and Contemporary Value of the Distinctive Features of Chinese Civilization" (CPC Leadership Group, 2024).

5.2 Technisierung

There are many ways to discuss the story of Western modernization, its successful struggle against ontological anxiety, the resulting deficit and unending quest to promote the advancement of scientific and technical progress. Regarding the term itself, one might speak of small-m modernization, for example, when a factory replaces old machinery with state-of-the-art technology and thus chooses to modernize. But then there is big-M Modernization which originates in German where *Modernisierung* has good company in similar words, some of them barely translatable: *Rationalisierung, Bürokratisierung, Demokratisierung, Säkularisierung, Mathematisierung, Mechanisierung, Medikalisierung, Cyborgisierung, Digitalisierung, Technisierung*. Each of these terms refers to a historical process of progressive advance along some trajectory. Depending on whether one likes the trajectory and where it is heading, these are viewed as human destiny and telos of history, or as corrosive of human life, as dangerous or decadent developments which need to be halted or reversed.

As for modernization one can critique and discuss it, for example, under the heading of "alienated reason" (*entfremdete Vernunft*)—as did Herbert Marcuse or, more explicitly, Gernot Böhme (Marcuse, 1964; Böhme & von Engelhardt, 1979; Böhme & Böhme, 1983). But even critics of the idea of progress cannot free themselves from the spell of the Needham question and from another kind of deficit that comes with it, namely a blindness of sorts to what the West missed out on. This blindness amounts to a deficit or confusion in our understanding of technique, technics, or technology as unfolding independently or outside the sphere of scientific progress, modernity and modernization.

According to the first of the three propositions above, China invented and developed many technologies, but according to the third it did not participate in technological modernization. These propositions interfere with and confound each other, as do the notions of *Technik* (technics, technique, technology) and *Technisierung* (technification, technological advance), as do the philosophy of technology and theories of technical development. There is on the one hand the acknowledgment of pre-modern and non-Western technical civilization, there is on the other hand the tendency to equate technical development as *Technisierung* with the ever-extending reach of modern science.

In order to disentangle and clarify this confusion, the term *Technisierung* begs to be critiqued. Does it make any sense at all to speak of technological progress as the world becoming progressively more technological? How are we to make sense of the claim that technical processes and devices pervade human existence ever more deeply? As a historical thesis, *Technisierung* seems to suggest that there is far more technology today than in the past: More and more of what was merely "natural" and thus presumably non-"technical" has been becoming technical. But clearly, this is wrong. The history of fire begins with flintstones, fire drills, or the harvesting and maintenance of lightning-induced fires. All of these are technical ways of harnessing fire, long before matches and lighters came along or that simple button on the gas grill. This is also true for our ways of feeding, dressing, housing ourselves. Even the history from cave paintings to video games is "technical" throughout.

Technisierung in all of these aspects of human life cannot mean that there is more and more technology, but might mean instead that the technologies become increasingly interdependent, requiring increasingly demanding systems for their maintenance. The simple procedures for killing animals and cooking their meat over a fire appear to be more isolated than a system of food production that renders a precooked meal in the supermarket and a microwave for heating it. But again, there is an obvious objection to this construction of the storyline: Today, I buy my precooked meal and in a matter of minutes I have eaten a substantial meal. In earlier days, the practice of hunting came with a whole way of life as tribes moved along with herds, setting up camps with tents made from leather, and the whole community, its fortunes and rituals are caught up in toolmaking, hunting, cleaning hides, preparing meals, conserving meat. That community's livelihood was determined by the technical requirements of the hunt, needing to be aligned with the movements of their prey. In light of this, can we really say that we are more firmly in the grip of our technologies and more dependent on them than were our ancestors?

Indeed, this objection gives rise to another, exactly inverted notion of *Technisierung*, according to which it is associated with notions of relief or release (*Entlastung*) (Gehlen, 1980). Technology promises to make our lives easier: As our world becomes more *technisiert* or pervasively technological, the less we need to do to sustain our livelihood, and the more we delegate to machines. But again, framed as a historical thesis, this sounds odd.[3] Contemporary AI represents a whole new dimension of *Entlastung* by releasing us of tedious cognitive labor, but it is also discussed as an extremely demanding technology: we have to defer to it, adapt and become attuned to it, we have to properly prompt and civilize it. AI systems share with wineries and assembly lines that they institute a form of life or technology game (Winner, 1988; Coeckelbergh, 2018). Once the *Entlastung* by way of AI is understood as a division of cognitive labor it proves to be not unlike the division of manual labor, comparable to the automatic loom and its pervasive impact initially on nineteenth century weavers, and still today on child laborers as well as a renaissance of artisan weaving practices. Are we really to believe that we are witnessing a linear development, a relentless process of *Technisierung*?

In his book *Invasive Technification* Gernot Böhme is also struggling with the difficulty of conceptualizing *Technisierung*. One of his paradigm examples of *Technisierung* provides another cue (Böhme, 2012). Consider the history of

[3] In the work of Edmund Husserl and Hans Blumenberg one also finds a notion of *Technisierung* that is not tied to a historical process at all, or only indirectly so: When we do not approach the world spontaneously or originally but use routines, techniques, or technologies we thereby render the world in a technical manner, we technologize the object of our experience, prepare and take things as tokens that play a functional role in processes of understanding or intervention. As Marcel Siegler has pointed out in an unpublished manuscript, it therefore remains unclear whether *Technisierung* has any specific role to play especially in regard to Husserl's project of analyzing the transformation effected by modern (technologized) science of the world of lived experience (Husserl, 1970; Blumenberg, 1981).

prosthetics. There is nothing novel about prostheses as such. They have been around ever since humans used sticks to support themselves. Over time, prostheses have crept up on us, no longer respect the boundary between a frail and vulnerable body on the one side, the technologies for supporting or propping up that body on the other. The prostheses nowadays pervade the body in the form of pacemakers or anti-depressants, they impose norms of normal, healthy, well-adapted living.[4] But again, is it really the case that prostheses insinuate themselves ever more seamlessly, effortlessly, or seductively into our pursuits, or is there perhaps something else going on here?

It has always been necessary to adapt the prostheses to the requirements of the human body and *vice versa*—the notion of uni-directional insinuation does not capture this mutual adaptation. This is true also for the biomechanically most advanced prostheses that promise seamless integration within a person's body-scheme, providing sensory feedback and the like. Martina Heßler shows that this integration has to be achieved always anew, it requires training, periods of adjustment, fine-tuning (Hessler, 2025). Each prosthetic technology poses problems of its own, including painful frustration and failure when humans prove incapable of controlling a device. Under the general heading of the "human-machine interface" the similarities and specific differences of such learning processes become interesting for the philosophy and cultural studies of any and all technology: How can and do we live our lives in and through a world of things?

What, then, remains in all of this of the overall tendency and trend that Böhme identifies and critiques as "*pervasive Technisierung*" (Fuchs, 2025)? Upon closer scrutiny it is not at all a trend of natural things becoming more technical. Instead, technical things are becoming more scientific. *Technisierung* is parasitic upon more and more of the world becoming amenable to scientific analysis. The creation of new and better prostheses is nowadays no longer a matter of trial and error or tinkering on the basis of common-sense knowledge. Instead, it pursues strategies of "rational design" and scientific understanding of human anatomy and physiology. This scientific understanding offers new interfaces, new notions of how and where prosthetic devices can latch on, intervene in causal processes, or support the work of human organs. And though it is impossible to make sense of *Technisierung* as a progressive advance along some trajectory regarding the pervasive significance of technologies, it is easy enough to speak of the progressive advance of science regarding its reach and depth. And when science expands its reach and improves its capacity to predictively control our life-world in ever more detailed ways, this affords new opportunities also for technical control.

Martin Heidegger drew attention to this in his discussion of the *Question Concerning Technology* (Heidegger, 1977): Technology is and has always been just

[4] Daria Bylieva is arguing similarly about digital technologies that more than any other they insinuate themselves into our ways of thinking, feeling, acting (Bylieva, 2023; Bylieva & Nordmann, 2023). And again, this seems quite comparable to the ways in which, for example, the assembly line or a world of plastics can produce a rhythm of life that reaches into the substrate of dreams and desires.

technology, disclosing (*entbergen*) causal features of the world. Modern technology is significantly different only in that it unfolds within the worldview of modern science which treats the whole of nature as a resource for technical exploitation: Technical modernization thus derives from or follows upon scientific modernization which seeks to smooth the path from prediction to control. To be sure, technical modernization might go by the name of *Technisierung* but at bottom it is *Verwissenschaftlichung* (scientification or scientization). Modern science and with it a specific form of rationality invasively pervades every aspect of life, and certain critiques of invasive technologies are really directed at a *wissenschaftliche Weltauffassung* which transforms how we think of ourselves and of the social and natural world. According to Max Weber and Martin Heidegger the basic conviction and program of modern science is "that one can, in principle, master all things by calculation" (Weber, 1946, p. 139).

Technological development seeks out, explores, inhabit pockets of reality that happen to yield to technical control—it does not require a principle, intellectual conviction, or progressive program. All technological systems and devices have in common that that they seek to establish a felicitous interplay of people and things—they settle at least temporarily on a way of doing things or a form of life (Winner, 1988). It is thus that life in the modern world is structured by its technologies neither more nor less than life in the medieval, ancient, or prehistoric world. But many of the technologies of the modern world are imbued by the progressive spirit or telos of modern science that seeks to render every aspect of life subject to causal analysis and predictive understanding.

And so we now have technical prostheses to regulate our heartbeat where all we had before were soothing words or the pills of healers. This is *Invasive Technification* or, rather, *Invasive Verwissenschaftlichung [Scientification]*: the healer used techniques as well, but only the pace-maker moves along the progressive trajectory of science. If the purpose is to regulate the heartbeat, technological development benefits greatly from the progress of medical and materials science. Even without valorizing "progress" but taking it only as a continuous advance along some trajectory, one can speak in such cases of the sciences enabling technological progress.

But by far not all technology develops progressively or keeps developing at all. André Leroi-Gourhan pointed out that certain problems of making stone-tools were conclusively solved in prehistoric times—there is a certain way of doing things and that's it, still the way of doing it thousands of years later (Guchet, 2024). With reference to Oswald Spengler, Ludwig Wittgenstein asked whether progress runs on and on, like a sausage, or comes to an end like a sausage as well (Wittgenstein, 2023, S. 23): Violin-making progressed for a while but the development peaked around the time of Stradivarius—since then, musically speaking, violins didn't get better anymore. And then, of course, there is Kafka's *Sorge des Hausvaters*, that is, the father's anxiety that some silly and senseless childish contraption will be around and survive him, indifferent to the telos of modernization (Nordmann, 2024). All of this is further testimony to the diagnosis that there is no general historical tendency or continuous advance of *Technisierung*. If the term is used nonetheless, it signifies that some fields of technology and technical practice—such as regulating the heartbeat—have benefitted from modern science and thereby advance a scientific way of doing and understanding things. In the case of violins, while the violins are not getting better, today's violin-makers ascertain the quality of sound not by hearing alone but also with the help of science-based measurement techniques—and one can interpret this as a *Technisierung* of the perceptual apparatus that relies no longer on highly cultivated "natural" hearing alone.[5]

5.3 Innerwordly Asceticism

Bewitched by the idea of a supposed historical tendency of things becoming ever more technical, one might take the history of prosthetics as a template for *Invasive Technification*, that is, for the "technological transformation" of humanity or the human life-world (Fuchs, 2025). With prostheses moving inside the body, becoming ever more closely aligned with physiology and biomechanics, it appears that humans are invaded by technology, taken over from within, fusing with technology also in their modes of communication, self-optimization, and perception. Another view is offered here: This chapter in the history of prosthetics is part of another story, namely that of "invasive scientization [*Verwissenschaftlichung*]", "rationalization" (quantification, mathematization), and modern culture. *Technisierung* is not a matter of this or that technical device turning the pristine natural human being into something that is only now deeply implicated within technical systems. Instead, it is about humanity implicated in a cultural movement that valorizes the progressive fusion of modern science, modern rationality, and capitalist modes of production and consumption. The critique of the very idea of *Technisierung* should now be complemented by a cultural diagnosis: What produces the seductive, seemingly inescapable conflation of *Technisierung* (technification) and *Verwissenschaftlichung* (scientization)?

Consider Donna Haraway's *Cyborg Manifesto* of 1985 in which she refers to herself as a cyborg. She explains that in post-World War II America, people became cyborgs not because "cyborgization" or *Technisierung* was creeping up on them, but because their modern scientific, political, and technological culture turned them into cyborgs. No one implanted a device into her body, she did not model herself after androgynous robots: She found herself a cyborg because she was born as a child of Sputnik and thus of the Space Age (Haraway, 1991, p. 173). Her biological parents gave her life, the Space Age gave her the world to grow up in and become a person.

[5] And where does the good old tuning fork fit in? It calibrates the objects of "natural" hearing through resonance—which is itself just one of the many phenomena that have been swept up, conceptually transformed, and technically operationalized through the advancement of science (Feldmann, 1997). The tuning fork serves as a technical as well as conceptual mediator between the human ear—as understood by Hermann von Helmholtz, for example—and the production of consonance, that is, between the ear as the "natural" apparatus and other instruments. Critical of Helmholtz, Matthias Rieger reads this as a chapter in the history of *Technisierung* (Rieger, 2006).

Before Sputnik, Catholic midwestern girls did not become Marxist biologists, but in the age of Sputnik the whole nation, boys and girls included, was recruited to the race for scientific and technical dominance of the United States. This implied new opportunities, new ways of becoming a person, namely a cybernetic organism: In this new world, people consider themselves part human and part machine, asking what makes personal and social beings superior or deficient to the machine. In earlier times, Catholic midwestern girls considered themselves part animal and part divine, asking whether the union of body and soul is in the image of God.

One becomes a cyborg by being implicated in a historical development which ties the fate of the modern world to scientific progress, rationalization, and capitalist modes of production and consumption. It is this peculiar juncture that in histories of modernization becomes salient twice: first at the beginning, in the age of Bacon with the Western spirit of modernization; and then again today, at the end of the Baconian age and a disenchantment with the modern world. Both moments need to be considered, beginning with the moment that was valorized by Needham who himself referred it back to Galileo's mathematized hypotheses about nature and Bacon's great renewal in the name of experimental science and the achievements of technology. It was there and then that the origin of modern science was to have made an enormous, seemingly disproportionate difference, transforming technology and industry, producing an idea of progress as an unending quest and hunger for the infinite.

This moment was theorized most famously by way of Max Weber's thesis about the Protestant ethic and the spirit of capitalism (Weber, 1930), coupled with Robert Merton's complementary thesis about puritanism and modern science (Merton, 1970).[6] This complementarity was prepared by Weber's essay "Science as a Vocation" (Weber, 1946). It can be summarized in a few words: As suggested above, mathematized hypotheses about nature respond to a peculiar ontological anxiety, namely the fear that we cannot truly know reality but only the constructions of our own minds. Seeking assurance that they are securely grounded in reality, scientists have to collect evidence, and they do so by casting out mathematically structured hypotheses, hoping that with these nets they capture predictively robust regularities. The collected evidence is never quite sufficient to provide the needed assurance, and the methodological posture adopted by modern science therefore maintains that hypotheses are always just this: more or less well-confirmed hypotheses. Similarly, the Protestant ethic also seeks assurance in the absence of certain knowledge. Am I among God's chosen people, will my soul be saved? The productive yield of one's methodical conduct of life may serve as evidence for a state of grace, but this evidence is never sufficient, can never serve as proof but requires the relentless

[6] There is a parallel track to this account: Alexandre Koyré, Mary Poovey, Lorraine Daston, and Ian Hacking considered the invention of the modern fact which owes to a fractured cosmos that is to be reassembled by way of universal laws. The methodological challenge of assembling and accounting for countless isolated facts arises together with the economic practice of book-keeping and gives rise to the problem of induction—a whole new philosophical problem and another expression of ontological anxiety (see, e.g., Koyré, 1968; Poovey, 1998; Hacking, 2002, pp. 12–13).

continuation of one's effort. Often, the "productive yield of methodical conduct" refers to the systematic generation of economic gain and thus to modern capitalism, but it also refers to the laboratory procedures of hypothesis testing and thus to the pursuit of unattainable truth. And "innerworldly asceticism" designates the methodical pursuit not only of capitalist accumulation but also of scientific knowledge (Weber, 1930, p. 154). It refers to a kind of selfless dedication, neither interested in personal gain, nor seeking salvation in some transcendent otherworldly realm, but tending to and rigorously accounting for worldly things.

Here, then, is the reason why Needham's minor matter is said to be of such great consequence: It is a historical fact that knowledge production by means of mathematized hypotheses developed initially only in the West, and this perfectly contingent fact signifies that methodical accumulation, progressive rationalization, mastery through calculation, science-based technology became a dominant cultural and economic force primarily in the West. The peculiar need to have evidence for one's standing in the world and for one's knowledge of reality is thus responsible for the supposed cultural deficit of nations that do not have that need and therefore did not adopt the habit of systematic accumulation and progressive expansion of science-based technological power. The stance of innerworldly asceticism proved immensely productive and it appears to be a necessary precondition for the emergence of modern science. But as one can learn from Needham, it is not a necessary condition for technological development, neither for pre-modern Chinese technology, nor for technoscientific playfulness.

Modern science is inherently progressive and *Verwissenschaftlichung* [scientification] along with *Rationalisierung* [rationalization] may well be claimed as historical tendencies of the modern age in the West and the sphere of Western influence. They are rooted in an anxious restlessness that can be valorized as an awakening, rebirth, or Baconian grand renewal with its unbounded spirit of exploration and discovery. If one wants to pinpoint this as a parting of the ways between China and the West, this can be done by contrasting Song Yingxing's *Tiangong Kaiwu* (1637) and Francis Bacon's *Novum Organum* (1620). Both works are grounded in a technologically advanced culture, but only Bacon embraces a vision of scientific progress, attended by the expansion of technical power.

5.4 A Parting of the Ways

They surely did not know of another, and yet they had a lot in common. Francis Bacon and Song Yingxing both embarked on an encyclopedic project. They broke with intellectual tradition in that they refused to engage in what they considered fruitless scholastic debate, referring instead to the achievement of technicians. Both view this as a collective exercise, they themselves can only make a beginning or show the way. Song Yingxing details how artisans circumscribe their world of 10,000 things. Certain ways of putting things together produce effects and it is through these effects that the world reveals itself to us. If only for this, the usual

English translation of the title has a distinctly Baconian ring: *The Exploitation of the Works of Nature*—in other words, exploitation not of the laws of nature but of its works (*opera*), many of which induced by human practice. Believing that one cannot learn from nature by contemplating or observing it, Bacon promoted the study of how the artisan or experimental composition of things [*opera*] exhibits the operations of nature. If Song calls for a larger research effort to catalogue and analyze the technical accomplishments of his day, Bacon conceives of academies that gather, catalogue, and investigate the works of nature.

If this is the shared understanding of Bacon and Song, the titles of Song's and Bacon's books also indicates a parting of the ways. Sinologist Dagmar Schäfer translates Song's *Tiangong Kaiwu* as *The Works of Heaven and the Inception of Things*. By choosing this title, Schäfer argues,

> Song intended to embrace a larger context than the simple documentation of craft procedures. He saw the revelation of universal principles within the performance of crafts and technology; it manifested a cosmological order.... Song's idea of what this interface between human and heavenly action implied differs fundamentally from our modern perception: man was not his focus. Instead he was interested in the knowledge displayed by the "works of heaven" and in the "inception of things." He believed that by describing the eighteen selected crafts man could learn the principles that would bring about order in the world. (Schäfer, 2011, pp. 17–18)

The works of the heavens and the works of nature reveal an ordered world as the things between heaven and earth prove beneficial not only for their utility but beneficial also for proper appreciation and artisanal emulation. In the words of Song:

> Of the tens of thousands of things and phenomena in the world, about half are of benefit to mankind and half not. A person who is endowed with intelligence and possesses a knowledge of the natural world is much respected by the multitude of ordinary folk. Yet there are those who...not knowing the measurement and care of cooking pots, would prefer to discourse emptily on ancient sacrificial vessels of Lü; when they paint they delight in depicting ghosts or monsters but scorn such common subjects as dogs or horses. Even if the talents of these people equal those of Kung-sun Ch'iao of Cheng and Chang Hua of Tsin, they are still, I say, unworthy of emulation! (Song, 1966, p. xi.)

Focusing on the large but limited number of "common subjects" in the world, Song's investigation of the works of nature situates these within the closed space of a cosmic order, and thus not within the abstract unbounded universe of modern science (Koyre, 1968). And so, for example, the woodcuts of pearl-diving, fishing, and boating show artisans on an open sea that is not really open but encloses them within a landscape of waves (Fig. 5.1).

In the "age of discovery" and colonial conquest, Francis Bacon ventures out into uncharted territory: *The Advancement of Learning, The New Organon*, or *The New Atlantis* are part of the *Grand Renewal* or *Instauratio Magna*. If for both Bacon and Song the inception of things and composition of works is the entry-point to an understanding of nature, it is only Bacon who considers composition as combinatorial, giving rise to an endless array of works. His skeptical position in Aphorism 4 of the *New Organon* provides the basis, drives, or feeds the unending quest expressed in Aphorism 48:

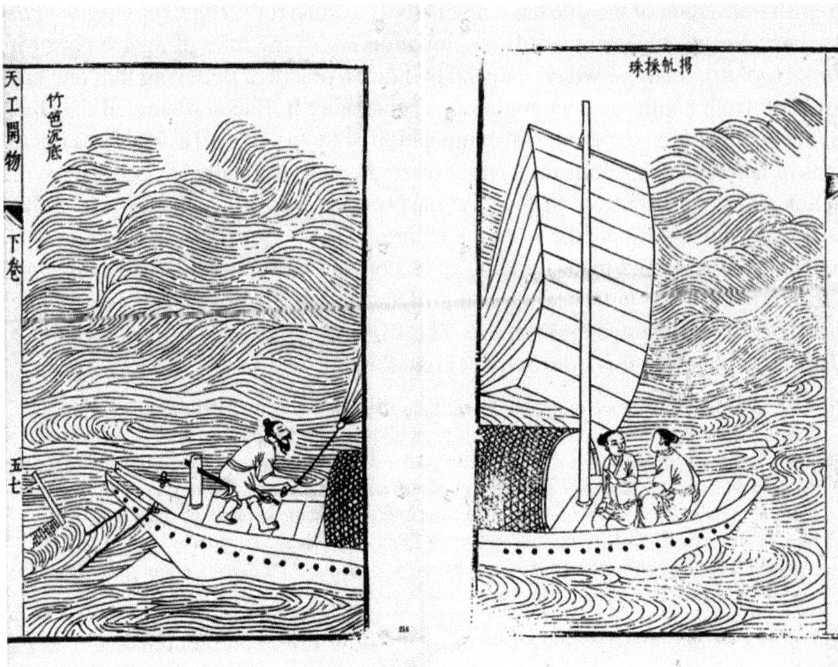

Fig. 5.1 Song Yingxing, *Tian gong Kaiwu*, 1959 (Facsimile of 1637 edition), Nanchang, compare Song, 1966, p. 297

> All man can do to achieve results [*opera*] is to bring natural bodies together and take them apart; Nature does the rest internally.
>
> The human understanding is ceaselessly active, and cannot stop or rest, and seeks to go further, but in vain. Therefore it is unthinkable that there is some boundary or farthest point of the world; it always appears, almost by necessity, that there is something beyond. (Bacon, 2000, aph. 4 and 48, pp. 33 and 44)

Bacon's ships stand ready to leave port and venture out to a great beyond. His ontological anxiety cannot be stilled since it is not enough for the human understanding to put things together and observe the effects produced by nature. Knowledge is accumulated as the ships bring back to the academicians of the *New Atlantis* many astounding items for inspection and analysis, but no matter how many phenomena they gather together, there is no limit to what they can produce. Bacon thus articulates the spirit of modern science that posits not a closed ordered world but an endless universe of which, perhaps, one can discover certain regularities here and there. This Baconian spirit does not confine itself to the certainties of the works between Heaven and Earth where things are effected in reliably artisanal ways. For Bacon, the works of nature and technological art are developed within an endless pursuit of novelty and originality (Fig. 5.2).

In this parting of the ways, Song Yingxing serves as proof of principle that there can be a Francis Bacon without the method or spirit of Galilean science, without

Fig. 5.2 Simon de Passe: Frontispiece in Francis Bacon's *Instauratio magna*, 1620

ontological anxiety, unending quests, and an unbounded universe. What remains is a philosopher of technology who carefully tends to the working order of things, principles of composition, measures and materials, technological forms of life, the felicitous interplay of people and things. Song treats technology as that which is "materially and intellectually a social appropriation of nature," proposing a program for its comprehensive investigation within a technological civilization.

5.5 Disenchantment

In the nineteenth and twentieth centuries, Karl Marx, Max Weber, Martin Heidegger, or Gernot Böhme formulated distinctly more somber accounts of Bacon's vision and the expansion of technical power. When Böhme proclaims "The End of the Baconian Age" (Böhme, 1993), he wants to show that innerworldy asceticm has run its course and with it what Max Weber has termed the "disenchantment of the world." Originally signifying the rationalization of a magical or mystical world, "disenchantment" no longer stands for Englightenment and liberation but now stands for our alienation from self and nature, for environmental degradation, pollution, resource depletion, and their attendant social conflicts (Böhme, 1993, pp. 24–26). Disillusioned by the Baconian project people today view scientific progress and modernization as profoundly ambivalent: Technological advance, in particular, is seen through the lens of ecological and social cost, or as a distortion of human conviviality. The prospective view at the dawn of the modern world shares with the retrospective critical assessment by a now reflexive "second modernity" (Beck et al., 1994) that both conceive the project of scientific modernization as the centerpiece for the ethos of rationalization and intellectualization as well as capitalist and consumerist exploitation of natural and human resources.

As he articulates his critique of the Baconian program Gernot Böhme defines it as a program for the reform and reorganization of modern science and society:

> Science as research, that is, as a collective system geared towards innovation—that is what Bacon demanded and what science really has become.... The House of Salomon is the concrete manifestation of Bacon's conviction that the human condition can be improved primarily through the development of science and technology.... He had thereby devised a type of society in which scientific knowledge has social authority and in which research makes up an significant part of public life.—This is the society we now have. (Böhme, 1993, pp. 11–13)

Böhme recognizes that today's disillusionment of that program is not concerned primarily with scientific knowledge or the organization of society, but instead with technology. In his chapter on "technical civilization" he therefore justifies his focus on technology in just this way:

> In our context science and technology are relevant as a single whole. This is grounded in the fact that the scientization [*Verwissenschaftlichung*] of our life-world consists not in adopting a scientific attitude in respect to our lives and not in everyday knowledge being transformed in scientific knowledge, but that...scientific products play an important role in our

> lives.... The basic tendency of rationalization and scientization consists in reconstructing our social and natural environment according to scientific insights. But surely there remains a difference between the symbolic representation of knowledge and its incorporation into structures of reality.... Both forms are representations of the same knowledge.... The question of what science is must be answered primarily through the consideration of its effects on human life. (Böhme, 1993, pp. 454–455)

To be sure, it makes sense to consider science in terms of its material technological implications, to view certain technologies as a vehicle for promoting and advancing scientization, and therefore to say that science is not an intellectual activity dedicated to theory-development but a technical program for expanding control over social and natural phenomena. But here we find Böhme oscillating in a curious way. For him "surely there is a difference" between science and technology and yet they are the same—and they are the same not because science adopts a technical or engineering stance which is what Heidegger and others have argued (Heidegger, 1977; see Forman, 2007). According to Böhme, they are the same because technology does as science does, and because technical devices and practices represent (scientific) knowledge: technology (re)constructs "our social and natural environment according to [its] scientific insights." This is a fruitful and productive way of looking at technology, but it also creates the predicament that he is no longer talking about a specific scientized trajectory of technological development but that he claims to be speaking about all technology as (re)construction and modeling the world, and—thus?—as science.

> It seems to me that science and technology [*Technik*] are a single interconnected project of humankind. The differences between them are differences of emphasis or orientation. For example, technology is by no means just the application of scientifc laws, instead it is a modelling of nature according to social functions. But such modelling takes place also in science itself. (Böhme, 1993, pp. 453–454)

What Böhme offers here is, on the one hand a social scientific conception of technology (*sozialwissenschaftler Technikbegriff*) (1993, p. 461), on the other hand a wildly implausible conflation of science and technology as a "single interconnected project of humankind"—with prehistoric origins of technology along with 400 years of Baconian science, conflating the technological civilization of 17th century China with the Western ethos of modernization, capitalism, and modern science.[7] Böhme hints at a way to reconcile these asynchronicities: "We are dealing with technology (*Technik*) just when there is a material as well as intellectual appropriation of nature or society" (p. 464). In other words, technology is technology only when it is like science, even before we can speak of modern science and society. Technology is thus a kind of proto-science and modern science consisted in the recognition of

[7] To be sure, the conflation or identification of scientific knowledge production and technical problem-solving makes sense in the context of a pragmatist epistemology which seeks to highlight an overarching conception of "inquiry." It does not make sense to project a scientific mind-set back onto all of technological development if one wants to develop a critique specifically of the Baconian program and of the technological effects of the *Verwissenschaftlichung* [scientization] of the modern world.

technology, e.g. mechanics, as science, and in the adoption of technical strategies, e.g. instruments and experiments (Böhme, 1993, p. 454). This construal of all technology as proto-science is based on an account of technology as "materially and intellectually a social appropriation" and functional modelling of nature — with the addition that "this is what science does as well." But what follows from this addition? If by metaphorical extrapolation from the sphere of science technologies can be understood as a form of modeling, and if modelling is part of scientific practice, does that make all technology (proto-)scientific? As intimated above, Song Yingxing in the absence of modernization and scientization also treats technology as that which is "materially and intellectually a social appropriation of nature."

This effort to reconstruct Böhme's philosophy of technology, as well as his conflation of *Technisierung* and scientization (*Verwissenschaftlichung*), finally leads to the heart of *Technisierung*: It is transformative in that it renders a previous condition or practice into something (more) technical as well as scientized, and thus further alienated or removed from self and nature:

> systematic arrays of material instruments...condition human relationships in quite a different way to mere rationalization; through their very existence, they set the conditions of possibility of the lives of individuals and societies. Driving a car is not simply a more efficient form of walking, to phone someone is not simply to speak to him at a distance, a sleeping pill is not simply a faster means of falling asleep, and the integration of our society via the internet is no mere rationalization of human interaction. As a systematic array of material means, technology does not simply leave the human relationships whose fulfilment it serves as they were, it transforms them structurally. And it is these structural changes which exceed the scope of concepts like rationalization and which ought to be referred to as technification in the true sense. (Böhme, 2012, section "Philosophy of Technology as Critique," compare 1993, p. 462)[8]

Also, according to Song Yingxing, the techniques, ethos, and attendant social order of pearl-fishing circumscribe a world of lived experience or "set the frame for what is possible by way of human and social life." But the full articulation and clarification of this technique does not follow a scientific spirit of rationalization and modernization, and does not go beyond it by way of a structural transformation in the direction of objectification and alienation. Böhme understands that he ultimately means scientification when he speaks of technification, and that he is therefore speaking not universally about technological developments but only about those that serve as vehicles for scientific progress. And yet, he projects the interconnection of science and technology in the project of modernity onto technology as manifestation of universal human practice.

Are there many different kinds of technology to focus on or only the overarching trajectory of technological development? In the end, unfortunately and all too simplistically, there is for Böhme only less and more *Technisierung* as we move from a world of culture into a technological civilization: "Technological civilization is as yet only a partially realized endeavour, something that we are in the process of producing" (Böhme, 2012, section "Civilization in the age of technoscience")

---

[8] Here, Böhme appears to draw on e.g. Husserl's use of the term *Technisierung* (see note 4 above).

And: "Moreover it is apparent that the worldwide advance of technological civilization has contributed substantially to the dissolution or collapse of traditional cultures.... The idea that a technological civilization will supply the culture of the new worlds of contemporary and future modernity is clearly an illusion." (Böhme, 2012, "Technological Civilization and Culture", see also Böhme, 1993). This is difficult to understand when every civilization is at its core also a technical civilization, and when the culture of any new or old world is also a technical culture.[9] And it does not get easier when Böhme seeks to illustrate the notion that "technology remains something distinct from culture" (Böhme, 2012, "Technological Civilization and Culture") by contrasting in in his book on *Atmospheric Architectures* two bookstores, one of them narrow, rustic, intimate, with books piled up among the wooden beams of an old house, a bookstore for booklovers, clearly. The other one is made of chrome, glass, steel, and neon lights, marble floors and islands for the display of bestsellers, a modernist temple for consumers, clearly (Böhme, 2017). The stores are distinguished not by their technical functionality but by their atmospheres and these "set the frame for what is possible by way of human and social life." For Böhme, the two bookstores are two worlds in that one frames the business of selling books as part of a technical civilization, and the other affords a space of cultural presence and lived experience. Another perspective would be more adequate and productive, however: The bookstores invite a critical comparison as contrasting technical solutions to the problems of selling books. They differ in what each of the bookstores affords and what they exclude: which one exhibits the spirit of capitalism, which one resists rationalist modernization? On this account, both are, albeit very different, expressions of technical world-making.

Gernot Böhme offers a sophisticated social scientific conception of technology but takes as the target of his critique not the differentiation among different modes of technological approach, but instead so-called *Technisierung* as such. In his influential conception of atmospheres, for instance, he appreciates them for what they afford, namely space for authentic lived bodily experience, an environment for sensing or feeling oneself in relation to conditions of livelihood and conviviality. It may be tempting to say that the romantic bookstore or an artisanal craft have atmosphere and a cold industrialized space or process do not—they simply do not allow us to be and to relate ourselves in any way at all. Böhme tends to that position and to treat atmospheres as essentially non-technical. At the same time, speaking of the arts of staging (*Inszenierungskünste*) he recognizes that architectural spaces can be atmospherically tuned just as one tunes a piano (Böhme, 2017). This does not appear to apply, though, to Wittgenstein's staging of a modernist building in Vienna (Wijdeveld, 1994). Discussing his design of the door-handle, Böhme notes the rejection of the ornamental, the universalist aspiration towards pure form that follows function, the negation of material particularity: Wittgenstein's design disguises not only any "communicative moment but human relations altogether. His

[9] To be sure, Böhme offers a sustained defense of his distinction between culture and civilization in the section on "Cultural resources for coping with technology" in *Invasive Technification* (Böhme, 2012).

door-handle does not derive in any way from a relation to the user, its form is not at all indebted to the human hand" (Böhme, 2002, p. 11). Instead of appreciating the specific atmosphere created by Wittgenstein's modernism—for example, his interest in human and material proportions—Böhme focuses on how it lacks atmosphere and not on how it creates one and not another.

Böhme's predicament is that he sees the creation and development of technologies through the glasses of modernization and rationalization, wrapped up with scientific progress and the spoils of capitalism. It complements Needham's predicament who sees the development of technologies in China and seems to expect that it would feed into the progressive rationalization and intellectualization effected by modern science. Both will be helped, so to speak, when the spectre of *Technisierung* as a historical tendency is lifted from stories about *homo faber*, about the works of technology and art, and the technical substrate of our forms of life. Finally, this also holds for Böhme's predicament in regard to "technical gadgetry" (Böhme, 2012). He appreciates the playful character of non-purposive "funny" devices but is finally compelled to place them within a narrative of entertainment and its progressive capitalist commercialization. More productively, instead, toys and other "funny" devices could figure prominently in reconstructions of technical cultures that exhibit and valorize tokens of technical power as a pleasurable end in itself—as evidenced by fireworks, paradigmatically (Nordmann, 2022). In analogy to artworks, "technical gadgetry" can be understood as material appropriation, transformation, or modelling of social reality, as conforming or resisting or indifferent to overarching tendencies of modernization, commercialization, and "scientization."

5.6 Misplaced Critique

Starting with the industrial age in the nineteenth century, the historical process of modernization and the intimate connection of science and capitalism has been discussed in terms of "alienation" (Karl Marx) and the "disenchantment of the world" (Max Weber). Weber and Marx simultaneously celebrate and critique rationalization and the explosive release of productive forces in capitalism. Proclaiming with Böhme the end of the age of Bacon is to foreground their critiques and those effects of the project of modernity which may have promoted its demise, highlighting that alienation and disenchantment are the most immediate products of "innerworldy asceticism." There is on the one hand the alienation of reason for which Kant prepared the ground (Böhme & von Engelhardt, 1979; Böhme & Böhme, 1983), there is on the other hand the disenchantment of the human body which has become progressively transformed into the physical body of (medical) science—*Körper* in German. Modern humans optimize, treat, sell their bodies and only when they get sick or die they are confronted not with having a body but with being their sensory body of lived experience (*Leib*). As the body of lived experience becomes disenchanted, turning into an object of natural, medical, social science, humans are losing what little has remained of "nature" in a scientifically and economically rationalized world—that last bit of nature is themselves, our being a sensory body (*Leib*). In order to save, cherish, and protect that last bit of nature Böhme argues for the "task of being a body" (Böhme, 2003).

This is an important story to tell as a critique of modernization and scientization, and it might undergird critiques of particular technologies. The story of *Technisierung* not as a historical trajectory but as the production of the simultaneous and successive worlds built by *homo faber* throughout the history of humankind, is totally different. And the Needham question, *ex negativo,* provides a vivid reminder of just how different technical world-making is from that very specific tendency of certain "emerging technologies" to be suffused with the scientific notions of progressive intellectualization.[10]

Language is how humans coordinate among each—often mediated by communication tools, recording devices, rhetorical techniques. Technology is how humans coordinate with things—often mediated by instructional manuals, research programs, verbal interactions. Language, in other words, is a kind of technology, and technology is a kind of language. Accordingly, the rejection of "technological civilization" makes no sense, just as little as would a rejection of "semiotic civilization." Nor does it make sense to say that we are not a technological civilization as yet—as if finally getting there would be the death-nail to human culture: There is no civilization, after all, that is not also a technological civilization. Whatever is wrong with the modern world, it is not technology or language *per se*. A critique of capitalism and consumerism, of industrialization or scientization makes more sense, questioning specific forms of life and how they contingently structure human relations, at the same time inviting us to imagine alternative trajectories. How to live, what technologies to promote, how to value, perhaps protect "nature"—these questions are negotiated in the spheres of politics and economics and sometimes of science. Our various technological forms of life reflect how, at one time or another, these debates were settled through our material and intellectual appropriation of nature.

Of course, the critics of *Technisierung* and our technological civilization are aware of all this. Technology becomes the target of their critique because any particular technological form of life is a material manifestation of our social organization—where the rubber meets the road. But surely, to critique a technological form of life is not to critique "technology." As one can clearly see from his books on the end of the age of Bacon and *Invasive Technification*, Gernot Böhme is a critic of technology only to the extent that he is a critic of *Verwissenschaftlichung*. Beyond that and despite this, he offers numerous clues for an appreciation of the creative life of *homo faber*.

Unfortunately, for Böhme as for many other German philosophers, cultural critique overshadows the beginnings for a coherent and comprehensive philosophy of

[10] Paul Forman provides a poignant diagnosis of the predicament: For the critical theorists of technology and culture, when they heard the word "technology" they thought "science" and thereby invisibilized technology. In recent decades and the age of technoscience, when one speaks of "science" one means "technology." This reversal signals a profound and ambivalent cultural shift which opens to view the questions concerning technology (Forman, 2007).

technology, one that puts all the pieces together. Needham's predicament poses a similar challenge. Fixated on the "Needham question" one becomes blindsided by specifically "modern technology" that owes its existence to the innerworldly asceticism of a scientific world-view as it arose in the West.[11] What might a philosophy of technology look like which takes as its point of departure fireworks and the piano, technical gadgetary or the technical forms of life in seventeenth-century China—one that takes all technical devices, techniques, and socio-technical systems as materially and intellectually appropriated nature and as modelling, ritualization, institutionalization of social practices? For this an aesthetic of atmospheres might need to be complemented by an aesthetics of felicitous accomplishment (*Ästhetik des Gelingens*)—the achievement of a happy interplay of things in a funny and serious work, the achievement of an interplay of people and things in at all times technologically constituted forms of life. If an aesthetics of atmospheres provides criteria for the suitability of a technically designed space for human well-being and conviviality, the aesthetics of accomplishment probes the non-violent, ideally frictionless working of a socio-technical system, the qualities of felicitous interplay.[12] As such it tends to principles of composition, the grammar and working order of people and things, thus contributing to a political philosophy of technology that reflects from within the reality of different political cultures (Mitcham, 2024).

References

Bacon, F. (2000). *The New Organon*. Cambridge University Press.

Beck, U., Giddens, A., & Lash, S. (1994). *Reflexive modernization: Politics, tradition and aesthetics in the modern social order*. Stanford University Press.

Blumenberg, H. (1981). Lebenswelt und Technisierung unter Aspekten der Phänomenologie. In H. Blumenberg (Ed.), *Wirklichkeiten in denen wir leben* (pp. 7–54). Reclam.

Böhme, G. (1993). *Am Ende des Baconschen Zeitalters. Studien zur Wissenschaftsentwicklung*. Suhrkamp.

Böhme, G. (2002). Wittgenstein im Begriff. In U. Kleefisch-Jobst & I. Flagge (Eds.), *Architektur zum Anfassen: FSB Greifen und Griffe* (pp. 9–14). Anabas.

Böhme, G. (2003). *Leibsein als Aufgabe: Leibphilosophie in pragmatischer Hinsicht*. Graue Edition.

Böhme, G. (2012). *Invasive technification: Critical essays in the philosophy of technology*. Bloomsbury.

Böhme, G. (2017). *Atmospheric architectures: The aesthetics of felt spaces*. Bloomsbury.

[11] Even as he seeks to consider Chinese technology "on its own terms," Needham finds odd that a technological civilization does not embark on the pathway of modern science. As was shown by Wu Wei and Mikael Hård, he did not shed the Western categories that dominate Chinese historiography of technology still today: the keywords are "invention," "revolution," "industrialization," and "modernization" as well as a substantivized "technology" with its singular *logos*. The present paper does not fully escape the paradox (Wu & Hård, 2020).

[12] Taken together, these interpretive efforts belong with emerging discussions of the hermeneutics of technology (for example, Kudina, 2024 or Grunwald et al., 2023).

Böhme, G., & Böhme, H. (1983). *Das Andere der Vernunft. Zur Entwicklung von Rationalitätsstrukturen am Beispiel Kants*. Suhrkamp.
Böhme, G., & von Engelhardt, M. (1979). *Entfremdete Wissenschaft*. Suhrkamp.
Bylieva, D. (2023). A semiverse of games. In D. Bylieva & A. Nordmann (Eds.), *The world of games: Technologies for experimenting, thinking, learning* (pp. 18–26). Springer. https://doi.org/10.1007/978-3-031-48020-1_2
Bylieva, D., & Nordmann, A. (2023). Entering the world technologically: Early encounters. *Technologos, 4*, 34–47. https://doi.org/10.15593/perm.kipf/2023.4.03
Carrier, M. (2011). 'Knowledge is power,' or how to capture the relationship between science and technoscience. In A. Nordmann, H. Radder, & G. Schiemann (Eds.), *Science transformed? Debating claims of an epochal break* (pp. 43–53). University of Pittsburgh Press.
Cavell, S. (1979). *The claim of reason: Wittgenstein, skepticism, morality, and tragedy*. Oxford University Press.
Coeckelbergh, M. (2018). Technology games. *Science and Engineering Ethics, 24*, 1503–1519.
CPC Leadership Group of the Chinese Academy of Social Sciences. (2024, January–February). Understanding the historical significance and contemporary value of the distinctive features of Chinese civilization. English edition of article in *Qiushi Journal*. Retrieved February 3, 2025, from http://en.qstheory.cn/2024-03/11/c_969196.htm
Feldmann, H. (1997). History of the tuning fork. I: Invention of the tuning fork, its course in music and natural sciences. Pictures from the history of otorhinolaryngology, presented by instruments from the collection of the Ingolstadt German Medical History Museum. *Laryngo-Rhino-Otologie, 76*(2), 116–122.
Forman, P. (2007). The primacy of science in modernity, of technology in postmodernity, and of ideology in the history of technology. *History and Technology, 23*, 1–152.
Fuchs, T. (2025). Lässt sich der Mensch verbessern? In R. Böhme, U. Gahlings, D. Mersch, A. Nordmann, & G. Schiemann (Eds.), *Philosophie als Kritik und Lebenspraxis: Zur Philosophie Gernot Böhmes*. Herder.
Gehlen, A. (1980). *Man in the age of technology*. Columbia University Press.
Grunwald, A., Nordmann, A., & Sand, M. (Eds.). (2023). *Hermeneutics, history, and technology: The call of the future*. Routledge.
Guchet, X. (2024). Leroi-Gourhan and the object of technology. *Technology and Language, 5*(2), 116–124.
Hacking, I. (2002). *Historical ontology*. Harvard University Press.
Haraway, D. (1991). A cyborg manifesto: Science, technology, and socialist-feminism in the late twentieth century. In D. Haraway (Ed.), *Simians, cyborgs and women: The reinvention of nature* (pp. 149–181). Routledge.
Heidegger, Martin (1977). "The question concerning technology" in M. Heidegger, The question concerning technology and other essays.: Garland, pp. 3–35.
Hessler, M. (2025). '…doch die Leiblichkeit seines Ichs erfahren'—Broken Body-Thinking und kaputte Cyborgs. In R. Böhme, U. Gahlings, D. Mersch, A. Nordmann, & G. Schiemann (Eds.), *Philosophie als Kritik und Lebenspraxis: Zur Philosophie Gernot Böhmes*. Herder.
Husserl, E. (1970). *The crisis of European sciences and transcendental phenomenology*. Northwestern University Press.
Koyré, A. (1968). *From the closed world to the infinite universe*. Johns Hopkins University Press.
Kudina, O. (2024). *Moral hermeneutics and technology: Making moral sense through human-technology-world relations*. Rowman and Littlefield.
Lin, J. Y. (1995). The Needham puzzle: Why the industrial revolution did not originate in China. *Economic Development and Cultural Change, 43*, 269–292.
Marcuse, H. (1964). *One-dimensional man: Studies in the ideology of advanced industrial society*. Beacon.
Merton, R. K. (1970). *Science, technology and society in 17th-century England*. Harper.
Mitcham, C. (2024). Brief for political philosophy of engineering and technology. *Nanoethics, 18*, 14. https://doi.org/10.1007/s11569-024-00463-8

Needham, J. (1954). *Science and civilisation in China* (Vol. 1). Cambridge University Press.

Needham, J. (1969). *The grand titration: Science and society in East and West* (p. 16).

Nordmann, A. (2022). Die schöne Technik der Verschwendung: Größte Kleinigkeiten. In S. Maasen & D. Atwood (Eds.), *Immanente Religion—Transzendente Technologie: Technologiediskurse und gesellschaftliche Grenzüberschreitungen* (pp. 269–285). Budrich.

Nordmann, A. (2024). Kafka's speculative technologies. *Technology and Language, 5*(3), 171–178.

Nordmann, A. (2025). Was heißt Technisierung? Fortschrittsgeschichten und Modernisierungsmythen. In R. Böhme, U. Gahlings, D. Mersch, A. Nordmann, & G. Schiemann (Eds.), *Philosophie als Kritik und Lebenspraxis: Zur Philosophie Gernot Böhmes*. Herder.

Poor, G. (2020). *The four great inventions: Technology, history, and nationalism.* University of Wisconsin-Madison. https://asset.library.wisc.edu/1711.dl/E6NLMS75AWUZR8X/R/file-cfee9.pdf

Poovey, M. (1998). *A history of the modern fact: Problems of knowledge in the sciences of wealth and society*. University of Chicago Press.

Rieger, M. (2006). *Helmholtz Musicus: Die Objektivierung der Musik im 19. Jahrhundert durch Helmholtz' Lehre von den Tonempfindungen*. Wissenschaftliche Buchgesellschaft.

Schäfer, D. (2011). *The crafting of the 10,000 things: Knowledge and technology in seventeenth-century China*. The University of Chicago Press.

Song, Y. (1966). Chinese Technology in the Seventeenth Century: T'ien-Kung K'ai-Wu, translated by E-tu Zen Sun. Pennsylvania State University Press.

Weber, M. (1930). *The protestant ethic and the spirit of capitalism*. Scribner.

Weber, M. (1946). Science as a vocation. In *From Max Weber: Essays in sociology* (pp. 129–156). Oxford University Press.

Wijdeveld, P. (1994). *Ludwig Wittgenstein: Architekt*. Wiese.

Winner, L. (1988). *The whale and the reactor*. University of Chicago Press.

Wittgenstein, L. (2023). *Movements of thought*. Rowman & Littlefield.

Wu, W., & Hård, M. (2020). History of modern Chinese technology: A review essay. *ICON: Journal of the International Committee for the History of Technology, 25*(1), 122–140.

Chapter 6
On Progress and Anxiety: A Lacanian Reading of Chinese Technology

Levi Checketts

Abstract Efforts have been made (in this book and elsewhere) to address Joseph Needham's question of why China did not modernize before the West. This paper takes a Lacanian psychoanalytic approach to the Needham Question and suggests that the key lies in how different cultures respond to the problem of anxiety. Western Christendom feels anxious about their impotence in the face of the Almighty and so creates technology in response. Chinese society, however, draws from different fonts for its worldview, and these yield different attitudes toward control or power, with only Confucianism offering an approximation for need to create technologies. However, much of this historical genealogy is irrelevant in the context of modern China, where anxiety is tied more to the existence and prominence of "the West" itself. Ultimately, anxiety is not something to be "cured," but China can avoid the temptation to follow the path of the West if it can adapt a free relation to its anxiety rather than seek to master it.

Keywords Jacques Lacan · Anxiety · China · The West · Technology · Daoism

6.1 Introduction: The Myths of the West and China

French psychoanalyst Jacques Lacan once wondered whether the Chinese had an unconscious. He had hoped to make a trip to China to inquire about this, expecting to discuss the matter with none other than Chairman Mao himself (Roudinesco, 1990, 545)! Lacan never got an answer to this question, but it is clear in his seminars, such as a lengthy discussion of Mencius in *Seminar XVIII* and long-lasting collaboration with Francois Cheng, that he was far more influenced by Chinese

L. Checketts (✉)
Department of Religion and Philosophy, Centre for Applied Ethics, Hong Kong Baptist University, Kowloon, Hong Kong
e-mail: checketts@hkbu.edu.hk

L. Checketts (ed.), *What Went Wrong with the West: Carl Mitcham's Critique of Western Philosophy of Technology*, Philosophy of Engineering and Technology 52, https://doi.org/10.1007/978-3-032-22214-5_6

culture and history than one expects of a mid-century French intellectual. It is fitting, then, that in recent years, Lacan has become a subject of interest among many Chinese philosophers and social theorists.

But it seems nobody has arrived at an answer to Lacan's question, at least not absolutely. Modern Chinese writers on Lacan talk about *modern* Chinese culture. And while some Chinese philosophy gets attention in Lacan's thought, it tends to be isolated from its broader social and historical context. The long cultural history of China does not get the same psychoanalytic treatment that Lacan is willing to give Western Christendom, where Sophocles, courtly love poetry, numerous saints including Augustine and Thomas Aquinas, and nearly every major philosopher, are put, so to speak, on Lacan's chaise lounge. Moreover, China in its modern articulation seems to suffer from a type of cultural amnesia. Indeed, we should note that China's entrée into the field of philosophy of technology is *as* a modern society, one that has developed modern attitudes toward technology due to its modernization, a theme I return to later in this chapter. Thus, drawing a comparison between China and "the West" in philosophy of technology itself elucidates large gaps in our concept of the field.

The provocative question Mitcham raises of "What went wrong with the West," an inversion of Joseph Needham's own question, raises the broader question of just what we mean by either "China" or "the West," and why and how these are different. Conceptually, academically, politically, and culturally we may speak of these as different, but history and geography complicate this. Why should North America be considered part of "the West" while China is part of the same land mass as most Western nations? Which elements of these cultures are distinct, given the exchanges of goods, ideas and peoples across Eurasia for millennia? Like the five Skandhas of Buddhism, if we try to identify either entity by one specific trait, we find it all falls apart. Europe has had far too many wars among its states, especially notable with the two world wars, to consider it a singular entity. China has likewise had its own succession of dynasties, political divisions and religious diversity to be a unified whole. We might thus ask, "Whose China? Which West?"

I contend that the question of how we articulate who and what China and the West are is important for articulating just how these different cultures have approached the problem of technology. For the sake of this chapter, I adhere to the Thomas Theorem—"If men define situations as real, they are real in their consequences" (Thomas & Thomas, 1928, 572). Contemporary global politics and the emergence of Chinese philosophy as an alternative to Western philosophy suggests that these divisions have been defined as real. But in line with Lacanian psychoanalysis, it is more probable that these distinctions are necessary primarily for shoring up our own identity. That is to say, they function as *symptom*, an external manifestation that we identify with (Lacan, 1975, 65; Žižek, 1989, 84). For the sake of this volume, I will focus on China's definition of itself over against the posit of the West and how this shifts Chinese approaches to technology. As such, I need first to examine what the West's relation to technology is from a psychoanalytic perspective, and how this relation conditions China's self-understanding in the modern period.

Mitcham suggests that the problem of the West's technocratic framework owes to an inherent fanaticism at root in Abrahamic faith (see also Mitcham, 2022). But certainly fanatics exist in Eastern traditions, and suppression of native religious beliefs is typical everywhere. Hindu nationalism today is a strong reminder of the former. Korean dynastic succession is illustrative of the latter, where Shamanism was prevalent in the Gogoryeo Dynasty before being suppressed by the Buddhist Silla Dynasty, and Buddhism itself was suppressed under the Neo-Confucian Joseon Dynasty. Thus, I take a different tack from Mitcham, and focus on the psychological structures typical of cultural-religious traditions. My overarching thesis, which I return to in the conclusion, is that the Chinese attitude toward technology has historically been fundamentally different from the West's because Chinese culture has been characterized by different types of anxiety from Christian anxiety, but in modernity, these divisions no longer hold. To demonstrate this, it is important to consider how anxiety plays out in the West and China. I thus first explain the connection in the West between *vociferated* anxiety and the pursuit of technology. I then consider China's philosophical past, attendant forms of anxiety, and why these do not lead to the same will-to-dominate that is present in the West. Finally, I turn to modern China, and its self-definition through symptomization of the West, resulting in a new anxiety and new relation to technology from the past.

6.2 The West: Masochistic Servants of the Almighty

The contemporary notion of the West is an invention of Enlightenment philosophers whose efforts to push for universal rationality entailed delineating their own philosophical tradition distinct from the rest of the World. In this account, "the West" consists of European (and colonial) cultures that claim some shared history traced back to the Roman Empire. In the Roman Empire, we find the confluence of Greek and Roman culture and philosophy, along with the spread of Christian religion. The influence of Celtic, Germanic or western Asian cultures on the formation of Europe is downplayed in this narrative, and the Fall of Rome is typically cast as the Romanization of the Barbarians, instead of the adulteration of Greco-Roman culture. Thus, our philosophical traditions ordinarily start with ancient Greece, and we speak of a Judeo-Christian civilization. Our universities, the site of much of this ideological production, have a long history of disciplinary overlap between theology and philosophy, which allows theologians to trace a lineage to Aristotle and Plato, and allows philosophers to positively cite Augustine and Aquinas. The university system itself, in which we typically count Oxford, Paris and Bologna as the founding schools, inscribes this tendency as we ignore the way Muslim scholars shaped medieval European centers of learning.

The *ex post facto* defining of a culture like the West is fraught with difficulties. However, it does seem plausible that ideologies do shape our encounter with the world and that they do have a bearing on particular attitudes toward technological development. Indeed, it is important to recall that philosophy of technology itself emerges from a recognition of the rapid advancement of industrial technologies in Europe and the US from the eighteenth to twentieth centuries and a corresponding impetus to create more technology, whether from an economic drive, a nationalistic sentiment, or even a religious belief. There *does* seem to be a thing like technological society, and it *does* seem to be co-extensive with Western modernity, so it *is* worth examining its cause.

Max Weber argued that the emergence of the "spirit of capitalism" was an effect of particular religious beliefs common in northern Europe and the US (2002). Lynn White, Jr., similarly argued that the devastation to the environment typical of industrialism has its roots in Christian beliefs about domination of nature (1967). Michel Foucault points out that many of the disciplinary techniques necessary for modern industry have their genesis in the practices of the Christian monastery (1995). Of course, philosophers of technology are familiar with these ideas. It is not my intention to either validate them or critique them, but rather to suggest there is reason to assume a connection between what one believes about ultimate things and the drive to technology. Theological beliefs are at least correlative in this broad account with technological society.

Let us go a step deeper, then, and bring Lacan into the picture. To put the question bluntly, why do Christians feel a need to dominate the earth? From a theological lens, Jacques Ellul contends that the need to tame and control nature through technological means is a result of the fall. Our expulsion from the paradisiacal Eden requires us to work and till the soil and creates an inherently conflictual relationship between humanity and the natural world (2022). Thus, to understand the Christian, and by extension the Western, worldview, we must understand a sense of inherited guilt through the original sin of our first parents, a gap between how things are and what they should be, and a desire to traverse this gap through technological means.

This gap constitutes, in psychoanalysis, our experience of anxiety. The fact that we do not experience holism, the feeling of incompletion, invokes anxiety as the experience of nothingness. Lacan supposes that all people experience anxiety (or at least, all people with an unconscious who have subjected themselves to the symbolic order). Anxiety is interconnected with desire, both of which relate to our experience of disunity with the Other. "The fear of desire is experienced as equivalent to anxiety…anxiety is the fear of what the Other in himself desires of the subject, this 'in himself' founded precisely on the ignorance of what is desired at the level of the Other" (Lacan, 2011, 308). In both anxiety and desire, we seek to fill in the lack. In desire, we seek to cover our lack with what we think the other desires as well, the "*Che Vuoi*?" (Lacan, 2019, 15). But desire itself is also ultimately insatiable. The inability to sate our desires leads to anxiety (Lacan, 2014, 316).

But the experience of anxiety and its form differs (Lacan, 2014, 291). Depending on the way by which we experience our inhibition, we will experience a different type of anxiety. Inhibition in the early stages of development can be the frustration of demand (anal) or of need (oral) (2014, 303). In later stages, anxiety is tied to the gap of our knowledge (scopic), the gap of our ability (phallic), and the gap of our morality (vociferated). The Christian in particular, Lacan notes, experiences a type of masochistic desire (2014, 220). In constituting the fallenness of humanity,

Christians seek expiation through pleasing their almighty God (2014, 307). But because God is almighty, and because we are not, the Christian experiences their anxiety as a type of masochism, a need to pursue moral rigorism and adhere most stringently to their faith as an attempt to expiate God's wrath (cf Checketts, 2025).

Christianity's masochistic anxiety results in an over-emphasis on good moral behavior. Ironically, the Christian belief that God alone saves them from their sins results in a greater effort to appease the wrath of God through merit. It was Søren Kierkegaard, after all, who first lit upon the importance of anxiety, which he directly tied to moral freedom (1980).[1] But ultimately, Christians believe in original sin, which means that from the beginning, we are already condemned. As Kierkegaard also points out, the only way out of this is through faith. And yet, this faith entails us holding onto the absurd position that we cannot merit our salvation, and yet we are saved. In the face of God as the Other, the Christian faces the ultimate uncertainty with regard to desire: an absent, silent Other who only desires our belief and nothing more. The path of faith which Kierkegaard encourages is one of learning "to be anxious will dance when the anxieties of finitude strike up the music and when the apprentices of finitude lose their minds and courage" (1980, 161–162). On a structural level, most Christians cannot live with this and so seek refuge in moralisms. Thus, as Weber points out, the Christian who cannot be assured of her own salvation seeks external validation of her status as elect (2002). In turn, then, Christianity, whether in the form of Catholic moral theology or Protestant moralism, has led to a *de facto* social system wherein Christians experience their desire for God's desire as unrequited, leading to an anxiety that propels them toward masochistic adherence to a Law that their theology officially teaches to be done away with in Christ.

The effect of this masochistic drive to "get it right" with God shapes the Christian relation to technology. Ellul posits the human control of technology cannot be separated from the reality of sin and the Fall. Compare this with the position of Thomas Malthus, who defends the industrial-capital system on theological grounds by contending that without the threat of starvation, we would not be propelled to develop advanced technological solutions to our ills. Both seemingly tie the need for technological conflict with nature to the curse of the fall, i.e., our own incompleteness, but whereas Ellul is critical of *technique* as a dominating, idolatrous social structure, Malthus thinks technology is one of the few blessings left us (1798, Chapter 19). Thus, technology holds this inherently diabolical nature, at least in much Protestant theology: it is a *necessary evil* we make an uneasy pact with because we otherwise only experience a hostile world. According to Weber, this particular development has its ties to Luther's sanctification of work itself as "calling." To

[1] One must not forget how significant the concept of anxiety is in Western philosophy following Kierkegaard either. Both Martin Heidegger and Jean-Paul Sartre give due attention to this phenomenon in their own *magna opera*, *Being and Time* and *Being and Nothingness*. The Christian existentialist theologian Paul Tillich likewise gives attention to this in his *Courage to Be*. While it is clear that Lacan is conversant and even influenced by this tradition, he also expresses discontent with the later philosophical articulation of anxiety. See Lacan (2019, 17).

declare that the manual pursuits that sustain life is what God calls us to is to say that excelling in crafts or production is to sacralize the otherwise profane business of ordinary life.[2]

However, this is not to say that Luther and Calvin deserve all the blame. Indeed, the great industrialists, so says Foucault, took notes from the Catholic tradition, which Weber himself sees as less capitalistic. The story Weber tells is that in Catholicism, the life of the merchant laity was a "lower" existence, and their economic pursuits inherently inferior to religious life. Thus, merchants were encouraged to "enjoy" the fruits of their work, sharing their produce with neighbors rather than hoarding it as pious Reform Christians must. But for Foucault, we must trace this masochistic relation to technology through religious life rather than lay life. He notes the model of the factory takes its disciplinary inspiration from the monastery (1995, 150). Monks' lives are characterized by *ora et labora*, praying and working. When not diligently praying, Christian monks were to be busy toiling, whether that be studying, copying, farming, or other work, all practices in self-mastery. Monasteries were thus the first institutions in the West to implement discipline, outlined, for example, in the *Rule of St Benedict* and implemented by a strict hierarchical order, and reinforced through disciplinary technologies like aesthetic practices and the mechanical clock.

Thus, we finally arrive at the most paradigmatic anxiety against our own impotence: the problem of time. Pioneering technology theorist Lewis Mumford noted "the clock, not steam-engine, is the key-machine of the modern industrial age" (1934, 14). Time is the singular unifying thread of all modern technologies: factories run on precision-timed machinery, telegrams are communication systems that transcend space and therefore save time, management science runs on principles of managing laborers' time, digital technologies carry out complex mathematical computations very quickly, high-speed transport delivers goods and persons long distances in short times, and so forth. The entire global capitalist system is rooted on principles of efficiency, i.e., that production is cheap, standardized and, most importantly, fast. Clichés like "time is money" or "too much time on one's hands" suggest the inherent tie between (the illusion of) mastery over time and broader technological domination.

Thus, the West's attitude toward technology, the efforts toward mastery, is characterized by the Christian's attitude toward time. In a cosmos where we have one life to live, a life characterized by our own shortcomings, and only a short amount of time to get everything right before we encounter final judgment and are sentenced

[2] Consider Mitcham and my discussions on the meaning of the scared and profane in relation to technology in Mitcham (2022) and Checketts (2022). Mitcham's position, sketched out 40 years prior, holds to what Weber would have considered a nominally Catholic position, i.e., that the technologies inherently belong to the domain of the profane. My position is that technologies can augment the sacred and even become sacred themselves through contact with the sacred, such as a saint's relic, in which case the technology no longer exists as a technology. Both of these positions are written and located in a typically Western framework, however, where the question of what is or isn't sacred is tied to a question of "getting it right," theologically speaking, though I think neither of our positions would be wholeheartedly considered either orthodox or heterodox.

to an eternity of happiness or suffering, we experience anxiety to please the almighty. This pushes us to seek to our own almightiness—moral rectitude, orthodox belief, and, ultimately, technological domination. The well-attested double sword of technological control and destruction is nothing less than a manifestation of Christian anxiety. Like systematic theology's attempt to know the ineffable God, or moral theology's attempt to prescribe good behavior to fallen humanity, the West's self-destructive pursuit of technology is nothing other than the structure typical of the psychology of the Western neurotic, characterized by a masochistic relation to God.

6.3 China: Searching for an Unconscious

China is a much harder beast to pin down. In the first place, the self-definition of the West was carried out as an intentional act of exclusion of other cultures; China's effort to define itself is a response to this question of the West. As Wu Guanjun notes, "In contemporary narratives of 'being Chinese'… 'the West' is often cast in the role of a violent force…. 'The West' thus becomes imaginable not only as the cause of Chinas' decline and the destruction of Chinese culture, but also as a force that needs to be accounted for, understood and studied" (2014, 110). A further problem is this: modern China, especially if we focus on the People's Republic of China, has a troubled relationship with its own history. In the Cultural Revolution, the Red Guard tried to root out all "regressive" elements of Chinese history, including desecrating the tomb of Confucius, destroying ancient texts, and ruining ancient temples. Since the "Rehabilitation" under Deng Xiaoping, China has made efforts to reclaim its own historical legacy, including greater support for Chinese Buddhism and formal Confucian education under current President Xi Jinping.

The problem of Chinese identity is not less complicated outside of the main land. Taiwan (the Republic of China) owes its own political history to the efforts to modernize China politically and philosophically during the Self-Strengthening Movement in the late nineteenth century. Hong Kong, Singapore and Macau, of course, have long colonial histories in which European culture, religion and politics have played outsized roles in shaping ethnic identity. The consequence ultimately is there is no unsullied Chinese legacy from which we can abstract some pure culture. Indeed, anecdotally, I note that several of my students from the mainland People's Republic of China have remarked to me that Hong Kong feels, in some ways, "more" Chinese to them than their home towns because of the way older cultural elements are maintained. Thus, the first real obstacle in asking why Chinese attitudes toward technology are different from Western ones lies in the fact that, despite China being articulated, both externally and internally, as a singular culture and history, it ironically lacks the same narrative cohesion that marks the nationally-diverse West.

All that said, most Sinology considers Chinese philosophical heritage to be strongly tied to Confucianism, Daoism and Buddhism—the three "master" schools—with other schools like Legalism and Mohism getting occasional mention (see Wang's contribution here). Let us consider each font of culture through the lens of anxiety.

6.3.1 *Buddhism and the Uncanny*

In contrast to Christian masochistic anxiety, Lacan ties Buddhist anxiety to illusive desire (2014, 222). Anxiety is always structured through desire and our inability to achieve that desire (2014, 41). The Christian desires to be almighty and so experiences anxiety through impotence and an effort to please the almighty. The Buddhist, on the other hand, desires to be free from the illusions of the world around them. Lacan calls this anxiety "scopic" anxiety, that is, anxiety related to the eye, or (mis) recognition (2014, 325). Thus, while the Christian experiences anxiety tied to failure to please the Other, the Buddhist plausibly experiences anxiety tied to misrecognizing the Other.

To demonstrate the notion of scopic anxiety, Lacan draws from Freud's essay on the Uncanny, which spends a great amount of time reflecting on the E.T.A. Hoffman story, "The Sandman," wherein the main character misrecognizes an automaton for a human and, at the climax of the story, mistakes his human fiancée for an automaton due to the automaton's human-like eyes (2014, 325). Fittingly, Japanese roboticist Masahiro Mori, himself a Zen Buddhist practitioner, is responsible for coining the popular use of "uncanny" in his hypothesis on "uncanny" anthropomorphic robots (2012). A more focused study on differing attitudes between Japanese and Westerners on artificial intelligence or robotics may further vindicate Lacan's position, but for the purpose of this chapter, suffice to say there is evidence of scopic anxiety in East Asian attitudes toward technology.

In Lacan's developmental schema, scopic anxiety precedes vociferated anxiety, the form of anxiety that pursue moral perfection due to our own failures, typified by the experience of a superego (2014, 295). The consequence, if this assessment is correct, is that Buddhists will have less of a sense of superego and likely a lesser sense of guilt compared to their Christian counterparts. On the other hand, they should be more likely to have anxiety tied to misrecognition, of misrecognizing themselves or the world around them. Perhaps unsurprisingly, then, there are psychological reports of Buddhists experiencing hallucinations while meditating or developing paranoid psychosis (Lindahl et al., 2013). If one's worldview emphasizes the need for correct perception, then it is inevitable that they will experience misperception in the consequent anxious experience of lack.

One can see in this vein how and why Buddhist monasteries differ from Christian ones. The Christian focus on stricture necessitates discipline and work; the Buddhist emphasis on seeing beyond illusion requires meditation and contemplation. Thus, the Christian monk attends to his external behavior in an effort to achieve internal perfection, while the Buddhist monk attends to his internal perception to transcend external deception. The Christian technologies of discipline use tools to shape the soul and mind; the Buddhist technologies of discipline use the mind to shape the body.

If Buddhism lacks a superego because it is characterized by scopic and not vociferated anxiety, this may account for the lack of interest in industrial progress from the Buddhist font. The Buddhist may not feel the same anxiety of wasting her energies because she is not concerned with almightiness but rather with illusion. Buddhism and Christianity thus differ on this central issue: for Christian culture, time is a threat and material is our ally, but for the Buddhist, material is an illusion and time is infinite. The Buddhist is not anxious that she will waste her one chance; she is anxious that she will remain trapped in the cycle of Samsara. Perhaps this is why Buddhism was easily supplanted by Neo-Confucianism in Korea and China—as an orientation to the world, it lacks the sort of anxiety that pushes one toward domination, which we will see is present in a form in Confucianism.

Of course, the Buddhist strand is only one of the three major "fonts" of Chinese culture. Daoism and Confucianism have a tense relationship with each other on the question of technology. As detailed in *Zhuangzi*, Daoism mistrusts "artful contrivances" as evidence of a "scheming mind" (*jixin* 機心literally "machine heart"), which is not "the proper residence of Dao." On the other hand, the followers of Confucius worry less about inner cultivation than about external behavior. Thus, as Mitcham points out in his own essay, cultivation of public order, the explicit aim of Confucian social teaching, promotes technological development for human ease. Daoism, with its focus on *wuwei*, distrusts interventionist approaches, so only those technologies that align with the natural order of things are acceptable.

6.3.2 The Castrated Confucian

Lacan never explored anxiety through Confucianism and Daoism as he does with Buddhism, but he does on occasion rather positively refer to their founding texts. Thus, the reading of these two other schools must be done through applying Lacan's pre-suppositions to the philosophical systems and the effects of these two philosophical traditions.

Wu Guanjun makes the strong case that the ancient masters of Confucianism and Daoism are perfectly compatible with Lacanianism. In his account, not only are the apparent tensions between Confucianism and Daoism, or between even Mencius and Xunzi, not true tensions, but even the appearance of these tensions illustrates why they underwrite Lacanianism. He suggests, for example, that Mencius's emphasis on *xin* 心 (heart/mind), Confucius's emphasis on *tian* 天 (the heavens), and Laozi's emphasis on *dao* 道 (the way) are all one in the same: each of these concepts signifies something other than what it might literally mean for us and so function as moving/metonymic signifiers (Wu, 2021, 207). In each case, the masters teach their followers to not be satisfied with the apparent order of things, but to always seek for the real that evades our structures. Thus, Wu reads positively Confucius's decision to leave off speaking as similar to Laozi's ineffability of the *dao* and Lacan's own prolonged silences in later life (ibid., 203). On Wu's reading, then, ancient Chinese philosophy is the pursuit of unstructured encounter of the Real that Lacan himself seeks (ibid. 206). This would, in turn, suggest a non-anxious encounter with the world: in escaping the confines of the symbolic order, one may pursue a free relation with lack rather than an anxious one.

In contrast to Wu's favorable reading, however, we should yet ask whether Confucianism and Daoism *as experienced* in China are actually supportive of this view. On this front, Wu and Lacan's generous readings of Mencius are undone by the social structures and strictures that are the legacy of Confucianism. Not a few authors have noted that Confucianism as a social system tends to enforce its structure through shame (see: Yang & Rosenblatt, 2001; Cheng & Rossner, 2023; Lu, 2018). One less enamored with an Orientalist fantasy might note that, Confucianism in a way parallels Christianity: regardless of what wisdom the radical teachings of the founder of the tradition raised, Master Kong's followers to this day have made his teachings into an impenetrable fortress; a veritable system of repression and impossible legalism.

The Confucian moral order is thus much like the Christian moral order. In *Civilization and Its Discontents*, Freud argues that civilization functions to instill repression in its members through functioning as a superego (2010). Although Freud is reflecting mainly on his own cultural context, this could just as easily describe a Confucian society as much as a Christian one: the economy of shame, social expectation, relation and demand plays a role of creating an outsized superego that represses all unrestricted eros. Thus, China and Korea have long tended to be rather sexually conservative, especially under Confucian influence. What makes a difference between Christianity and Confucianism, however, is that while Christianity focuses on the almightiness of the invisible God, Confucianism ties its moral expectations to human society, both the living around us and the dead gone before. Thus, while Christians worry about their worthiness before an almighty God, Confucians worry about their worthiness before their ancestors. Both systems function through vociferated anxiety—it is the voice (of God, of one's parents) that provokes anxiety.

Wu contends that the Chinese belief that "to receive the *full* amount of greatest joy…[one must achieve] the realization of being congruent with oneself…. This is probably why China has developed a rich and powerful ethics, but not capitalism" (2021, 211). In this passage, Wu focuses on Lacan's notion of *jouissance*, not anxiety, but the gap between Wu's contention and Lacan's position reveals where anxiety creeps in. The Confucian or Daoist presumption (according to Wu) is that one *can* align their psyche with the ineffable real. The position of Lacan is that all speaking subjects are *castrated* subjects—the process of adopting language entails losing our contact with what is real (Lacan, 2020). Thus, we perpetually feel the gap, the hole present in being which cannot be filled. This gap is responsible for desire (seeking wholeness), *jouissance* (going beyond fullness), and, of course, anxiety. The symbols of our culture frame our relation to this gap. Thus, the "rich and powerful ethics" Wu proudly touts are not the transgressive ethics of desire that Lacan espouses, but rather the pleasure-principle based law Lacan condemns in Aristotelian ethics (Lacan, 1997, 292; cf. Wu, 2021, 210); whatever may be said about the congruence of Lacanianism with ancient Chinese philosophy, it is clear that Confucianism's ethical foundation is essentially tied to the symbolic structure that has anxiety at its heart.

At this point, it is worth iterating how much of Chinese ethics of technology today takes its starting point from Confucianism. Wong and Wang's (2021) collection of essays, for example, outline guidelines and normative positions toward technology from a Confucian standpoint. Indeed, scholars like Pak Hang Wong (2020), Tom Xiaowei Wang (2021) and Qin Zhu (2020, 2023) have made it a goal to cement Confucian ethics of technology as a viable alternative to Western approaches. The same cannot truly be said of the other two major sources of Chinese philosophy. While there is some work on Buddhism and technology (Hershock, 2021; Hongladarom, 2020), there is virtually no attempt to make a Daoist ethics of technology (notable exception: Teschner & Tomasi, 2009). Confucianism presents itself as a rival source for ethics of technology against Western philosophical discourse because among these three traditions, it has the strongest force of repression and vociferated anxiety.

The main source of division between a deeply Confucian social order and deeply Christian one, and perhaps the root of the difference between China and the West with regard to technology, is the issue of almightiness. The surplus *jouissance* that characterizes capitalism, the excess repression of eros Marcuse similarly notes (1962), we recall is posited through Weber as needing to be rechanneled into one's labor—the innovative-investment paradigm typical of industrial capitalism and its general system, whether one calls that *technique*, technological society, capitalism, Western society, or what have you. But Confucianism is inherently backward-looking. The Confucian's anxiety is not tied to the expectation of a sadistic God who will punish, but rather the disapproval of their ancestors coming from behind. Thus, as Mitcham himself notes, the Chinese philosophy toward technology has typically been one of propriety—how to rule well or promote social harmony rather than one of conquest, preservation rather than domination.

6.3.3 *The Dao That Cannot Be Said*

Buddhism is more or less indifferent to technological progress, concerned as it is with *maya*. Confucianism approaches the question from the question of social order, necessitating the creation of technologies that will promote social harmony. Daoism, for its part, seems most congruent with Lacan's own views and simultaneously offers the strongest rejection of technology among the three master philosophies. To illustrate the Daoist attitude toward technology, consider the following passage in *Zhuangzi*:

> Zi-gong [one of Confucius's disciples] had been rambling in the south in Chu, and was returning to Jin. As he passed (a place) on the north of the Han, he saw an old man who was going to work on his vegetable garden. He had dug his channels, gone to the well, and was

> bringing from it in his arms a jar of water to pour into them. Toiling away, he expended a great deal of strength, but the result which he accomplished was very small. Zi-gong said to him, 'There is a contrivance here, by means of which a hundred plots of ground may be irrigated in one day. With the expenditure of a very little strength, the result accomplished is great. Would you, Master, not like (to try it)?' The gardener looked up at him, and said, 'How does it work?' Zi-gong said, 'It is a lever made of wood, heavy behind, and light in front. It raises the water as quickly as you could do with your hand, or as it bubbles over from a boiler. Its name is a shadoof.' The gardener put on an angry look, laughed, and said, 'I have heard from my teacher that, where there are ingenious contrivances (機械), there are sure to be subtle doings; and that, where there are subtle doings, there is sure to be a scheming mind (機心). But, when there is a scheming mind (機心) in the breast, its pure simplicity is impaired. When this pure simplicity is impaired, the spirit becomes unsettled, and the unsettled spirit is not the proper residence of the Dao (道). It is not that I do not know (the contrivance which you mention), but I should be ashamed to use it.' (Zhuangzi n.d., "Heaven and Earth" 11).

Zhunagzi's gardener, whom we are meant to read as possessing deep Daoist wisdom, confronts Confucius's disciple. Zi-gong's "solution" is understood to be an effort to circumvent the simple life, a "scheming" solution. In Lacanian terms, we might say it is an attempt to cover up one's lack. Thus, the Daoist approach advocates *not* trying to connive one's way out of the difficulties of life.

The attention Daoism gives to the seeming chaos of the world is manifest in *its* ethical orientation, especially the primary concept of *wuwei* (無為). Wu translates this notion as "non-action/whatever-action" (Wu, 2021, 210). The ambiguity of this translation is important—it conveys the greater sense of "going with the flow" rather than mere inaction. It is neither rule-bound nor transgressive, but rather adaptive to changing reality. Unsurprisingly, then, Lacan himself expressed most clear affinity for Daoism, especially finding congruence between the division between the symbolic order and the real depicted in the *Daodejing*'s opening line, "The Dao (道) that can be followed is not the eternal Dao (道). The name that can be named is not the eternal name. The nameless is the origin of heaven and earth, while naming is the origin of the myriad things. Therefore, always desireless, you see the mystery, ever desiring, you see the manifestations. These two are the same—when they appear they are named differently. This sameness is the mystery, mystery within mystery; the door to all marvels" (Laozi, n.d., 1; Cheng, 2002, 58; cf. Wu, 2021, 208).

Daoism's emphasis on paradox and the futility of effort parallels Lacanian psychoanalysis's approach to anxiety. The goal is not, despite what it may seem, to be *rid* of anxiety; the goal is to accept it and recognize it for what it is. Master discourses, such as ego psychology, consumeristic corporate strategy, religious fundamentalism, insist that *a* solution to the problem of anxiety exists and can be achieved, but Lacanian psychoanalysis notes this is merely the obsessional neurotic's attempt to stave off what cannot be ultimately eradicated, namely the emptiness of existence (Lacan, 1970) . The gap cannot be eradicated as long as we are linguistic subjects; thus, we must learn to be comfortable with the anxiety. Daoism emphasizes giving up one's efforts to achieve wholeness or control. The depiction of laughing master Laozi abandoning civilization on the back of a water buffalo represents such an attitude toward the gap. Daoism might be said to focus on sublimation, an escape or transgression against the symbolic order rather than conformity with it (Lacan, 1997, 322).

But because Daoism in itself does not have a strong ethical focus, nor a political orientation, nor even practical suggestions for how to live one's life, and because other philosophies promise a solution to anxiety (though they cannot deliver), Daoism is usually co-opted into another tradition and loses its potency. Thus neo-Confucianism, which maintains all of the political and ethical anxiety of its predecessor, co-opts Daoism as a metaphysical grounding, and Zen Buddhism takes its bearing toward emptiness from Daoism. Thus, like Wu's reading of Chinese philosophical masters, whatever we might say about the attitude Daoism advocates toward anxiety, the reality lived in Chinese history is domestication and structuralization resulting in a conventional and a strong symbolic order. Thus, whatever uneasy peace Daoism may offer to anxiety, by virtue of its focus on inaction, it is easily overtrod by structured systems demanding more. Sublimation is a lonely road; those who pursue it live outside the bounds of the symbolic order. Most of us cannot realistically pursue this mode of life, so more conventional forms of anxiety tend to triumph. In the case of China, Neo-Confucianism becomes the paradigmatic cultural and political expression, defining the historical philosophical orientation toward technology. Daoism, for its part, becomes the purview of either those seeking to escape the bounds of the symbolic order or those wielding it as a type of conventional, traditional religious attitude compatible with other philosophies.

6.3.4 Modern China and the West as Symptom

The ancient Chinese philosophical account of technology can only go so far. We are, after all, encountering traditions read through a romanticist lens, "through a glass darkly" as it were. There is no unbroken tradition nor even a pretending of such a thing as we consider the question today. Or, as Wu notes, "The narrative about 'the bursting rumble of guns and cannons that broke the peace of the night,' fantasmatically and retroactively allows people to imagine that there was a 'peaceful, harmonious China' before the intrusions of the West broke it. The existence of an essential Chinese civilization or 'Chineseness' only becomes imaginable once it was perceived as being deprived" (2014, 104–105). Modern China is ultimately a different entity, and anachronistic readings of cultural history can only sketch out some of the historic structures repressed within the Chinese unconscious. It is thus more interesting to consider how China's attempt to define itself in relation to the West has massively changed its attitude toward technology.

The notion of "progress" is a recent addition to Chinese language. The typical way of rendering this word, in the assorted meanings we tend to imply in English, is *jinbu* (進步). This word, commonly used today, seems to have gained currency in the Self-Strengthening Movement of the late nineteenth-century (Fröhlich, 2020). Prior to the twentieth-century, the idea of "moving forward" would have been seen as an absurdity in much the same fashion that Westerners use the term "backwards" in a pejorative sense (Vogelsang, 2020). In the Self-Strengthening Movement, Chinese intellectuals, such as US-educated first Chinese president Sun Yat-Sen, sought to overturn the "backwardness" of the Qing Dynastic legacy and modernize China, bringing the country "up to date" with the modern world. At this pivotal point in history, after numerous tense interactions with the West including the Opium Wars, the Boxer Rebellion and the Taiping Heavenly Kingdom, the worldly literati of Chinese society sought to remake the Middle Kingdom in the image of Western colonial powers.

The effect of this is non-negligible. Despite the Republic of China only maintaining power in the mainland from 1912 to 1949, the People's Republic of China was even more emphatic to be free of the trappings of China's history. Mao Zedong famously sought to root out the "four olds" (ideas, cultures, customs and habits) during the Cultural Revolution as being anti-Proletarian and positively feudalistic. In place of Chinese relics, the people should champion Newborn Socialist things. The ideology of German philosopher Karl Marx, already dominant in the Soviet Union north of China, should become the regnant philosophical orientation in place of the antiquated Chinese philosophies, not only economically, but also culturally and politically.

The Chinese notion of progress, however, has ushered in a new form of anxiety for China. By adopting a Western idea as pivotal for its modern identity, modern China has come to define itself *by* the West. The West stands for China as its *objet a*, that thing by which we identify ourselves (Lacan, 2014, 40). The West, for China, is an object of loathing and desire. The notion of "self-strengthening" insists on a need for an independent China—independent because by becoming like the West, it is now on the same playing field as the West. And so, *jinbu* becomes essential for the modern Chinese spirit: our values must be progressive, our economy must be progressive, our educational system must be progressive, our *technology* must be progressive. The American spirit of innovation, which *Zhuangzi* would denounce as *jixin* pure and simple, is today the stated goal of the Chinese National Natural Sciences Foundation and numerous funding projects, state-sponsored corporations, and 5-year economic plans.

Thus, it has become important for modern Chinese identity to reclaim ancient technological developments as authentically Chinese, including gunpowder, paper and the printing press. Western appropriation of Chinese technologies has always been toward domination. If the Chinese invented gunpowder, it was the Europeans who developed this technology's full potential, using it to conquer distant lands and create global empires. If the Chinese invented printing, Europeans took this technology to disseminate their own language across the world so that, even if more people natively learn to read Chinese, they must learn to read English if they wish to go outside of their borders. And thus, the need to claim these technologies as authentically Chinese reflects nothing short of a need to demonstrate the "progress" inherent in Chinese culture all along. It is a redefinition of Chinese history in terms of Western values, an anxiety tied to the gap between the West and China.

In the end, or rather, at present, the question of "what went wrong" has become irrelevant, not because it isn't an important question, but rather because China has adopted the same masochistic anxiety that characterizes the West. As China has taken the place of the fallen Soviet Union as the antipode to the United States in the global political order, it defines itself in connection to this paradigmatically Western nation. Whatever the unconscious structures defining Chinese culture and policy prior to modernization, the past century has been one of increasing need to identify with the West, which means sharing the same attitude toward technology. Ironically, then, while the West's transformation into Industrial capitalism owes its origin to anxiety in the face of the Almighty, China's form of this change owes to its anxiety in the face of the West.

6.4 Conclusion: Beyond Sense to the Dao

The attempt to understand "what went wrong" is a way of casting the past and particular historical developments in such a way as to vindicate the ideological claims of the capitalist, i.e., that the "West" is of a particular culture prioritizing technology as "progress" over the untamed world and that other "uncivilized" territories must be cultured. China's own negotiation of this narrative for the purposes of constructing a national narrative further supports this particular claim. But both attempts are efforts to side-skirt the reality of lack: Western violence to persons or environment (and let us not ignore Eastern violence, such as the imperialism of the Mongols or the Japanese) are realities that have produced long memories, which we seek to make sense of by integrating them into a broader narrative to avoid the fact that violence is a brute fact with no sense.

Violence escapes language because it is a deliberate destruction of the symbolic, a "traumatic confrontation" only experienced in the real (Lacan, 2014, 312). Language is an attempt to rein this in, to make sense of what is nonsensical, to provide a clear justification for why things happen, to control reality through sense-making. The violence of technological impositions have to be framed within a coherent symbolic order if they are to be sensible to us and not an instance of psychosis. Cutting off a mountain top to get the coal inside to power our power plants doesn't mean anything—it is nonsensical and, as Deleuze and Guattari suggest, nearly schizophrenic (2009).

The effort to "make sense" of it all is another such example of domination. Here, China has long been subject to "Western" violence—the need to define Buddhism, Daoism and Confucianism as philosophy or religion, distinctly European concepts, the need to compare civilization "progress" (a concept introduced to China from Marx), the establishment of a global order of nation states which compete against each other, etc. To ask "what went wrong" is already to be trapped in the iron cage, to be stuck in *Gestellung*. Any attempt to recover the past will be mistaken. What we should do is pose the more daring question: why does China want to be like the West? How can we transcend this instrumental rational approach to comparative philosophy? And how do we confront this anxiety bare faced rather than cowering behind "artful contrivances" (機械) which do not allow for the in-dwelling of the *Dao*?

References

Checketts, L. (2022). Christ and Technology in Dialogical Relation: Some Reflectons on the Technological Augmentation of the Sacred. In C. Mitcham, J. Grote, & L. Checketts (Eds.), *Theology and technology Vol. I: Essays in Christian analysis* (pp. 91–115). Wipf&Stock.

Checketts, L. (2025). A Lacanian analysis of Christian paradox in moral teaching and anxiety. In A. Chavoshian & S. J. E. Park (Eds.), *Anxiety in the era of uncertainty: An interdisciplinary reading of Lacan's seminars* (pp. 51–68). Wipf&Stock.

Cheng, F. (2002). Lacan in Chinese thought [拉康与中国思想]. In Y. Daiyun (Ed.), *Cross-culture dialogue 8* (pp. 54–65). Shanghai Culture Press.

Cheng, R., & Rossner, M. (2023). Unpacking shame and Confucian relationalism in Taiwanese restorative justice. *Asian Journal of Criminology, 18*, 209–230.

Deleuze, G., & Guattari, F. (2009). *Anti-Oedipus: Capitalism and schizophrenia* (R. Hurley, M. Seem, & H. R. Lane, Trans.). Penguin.

Ellul, J. (2022). Technique and the opening chapters of genesis. In C. Mitcham, J. Grote, & L. Checketts (Eds.), *Theology and technology Vol. II: Essays in Christian exegesis and historical theology* (pp. 27–41). Wipf&Stock.

Foucault, M. (1995). *Discipline and punish: The birth of the prison* (A. Sheridan, Trans.). Vintage Press.

Freud, S. (2010). *Civilization and its discontents* (J. Strachey, Trans.). W. W. Norton & Company.

Fröhlich, T. (2020). Progress, history and time in Chinese discourses after the 1890s. In T. Fröhlich & A. Schneider (Eds.), *Chinese visions of progress, 1895 to 1949* (pp. 1–40). Brill.

Hershock, P. (2021). *Buddhism and intelligent technology: Toward a more humane future*. Bloomsbury Academic.

Hongladarom, S. (2020). *The ethics of AI and robotics: A Buddhist viewpoint*. Lexington Books.

Kierkegaard, S. (1980). *The concept of anxiety: A simple psychologically orientating deliberation on the dogmatic issue of hereditary sin* (R. Thomte, Ed. & Trans.). Princeton University Press.

Lacan, J. (1970). *The seminar of Jacques Lacan XVII: Psychoanalysis upside down/the reverse side of psychoanalysis* (C. Gallagher, Trans.). LacaninIreland.com.

Lacan, J. (1975). *The seminar of Jacques Lacan XXII: R.S.I.* (C. Gallagher, Trans.). Online PDF edition at http://www.lacaninireland.com/web/wp-content/uploads/2010/06/RSI-Complete-With-Diagrams.pdf

Lacan, J. (1997). *The seminar of Jacques Lacan VII: The ethics of psychoanalysis* (J.-A. Miller, Ed. & D. Porter, Trans.). W. W. Norton & Company.

Lacan, J. (2011). *The seminar of Jacques Lacan IX: Identification* (C. Gallagher, Trans.). Lacan in Ireland.

Lacan, J. (2014). *The seminar of Jacques Lacan X: Anxiety* (J.-A. Miller, Ed. & A. Price, Trans.). Polity Press.

Lacan, J. (2019). *The seminar of Jacques Lacan VI: Desire and its interpretation* (J.-A. Miller, Ed. & B. Fink, Trans.). Polity Press.

Lacan, J. (2020). *The seminar of Jacques Lacan IV: The object relation* (J.-A. Miller, Ed. & A. Price, Trans.). Polity Press.

Laozi. (n.d.). *Daodejing* (J. Legge, Trans.). Chinese Text Project. https://ctext.org/dao-de-jing

Lindahl, J., Kaplan, C., Winget, E., & Britton, W. (2013). A phenomenology of meditation-induced light experiences: Traditional Buddhist and neurobiological perspectives. *Frontiers in Psychology, 4*, 973.

Lu, Y. (2018). Shame and the Confucian idea of *Yi* (righteousness). *International Philosophical Quarterly, 58*(1), 5–18.
Malthus, T. R. (1798). *An essay on the principle of population* as it affects the future improvement of society, with remarks on the speculations of Mr. Godwin, M. Condorcet, and Other Writers (1st ed.). J Johnson.
Marcuse, H. (1962). *Eros and civilization: A philosophical inquiry into Freud*. Vintage Books.
Mitcham, C. (2022). Theology and technology revisited. In C. Mitcham, J. Grote, & L. Checketts (Eds.), *Theology and technology Vol. I: Essays in Christian analysis* (pp. 116–131). Wipf&Stock.
Mumford, L. (1934). *Technics and civilization*. Routledge.
Roudinesco, E. (1990). *Jacques Lacan & Co.: A history of psychoanalysis in France, 1925–1985* (J. Mehlman, Trans.). University of Chicago.
Teschner, G., & Tomasi, A. (2009). Technological paradigm in ancient taoism. *Techné: Research in Philosophy and Technology, 13*(3), 190–205.
Thomas, W. I., & Thomas, D. S. (1928). *The child in America: Behavior problems and programs*. Alfred A. Knopf.
Vogelsang, K. (2020). The Chinese idea of 'progress'. In T. Fröhlich & A. Schneider (Eds.), *Chinese visions of progress, 1895 to 1949* (pp. 43–74). Brill.
Wang, T. X. (2021). Confucian ritual technicity and philosophy of technology. In P. Wong & T. X. Wang (Eds.), *Harmonious technology: A Confucian ethics of technology* (pp. 10–28). Routledge.
Weber, M. (2002). *The protestant ethic and the "spirit" of capitalism and other writings* (P. Baehr & G. C. Wells, Ed. & Trans.). Penguin Books.
White, L. (1967). The historical roots of our ecological crisis. *Science, 155*(3767), 1203–1207.
Wong, P. (2020). Why Confucianism matters for the ethics of technology. In S. Vallor (Ed.), *The Oxford handbook of philosophy of technology* (pp. 609–628). Oxford University Press.
Wong, P., & Wang, T. X. (Eds.). (2021). *Harmonious technology: A Confucian ethics of technology*. Routledge.
Wu, G. (2014). *The great dragon fantasy: A Lacanian analysis of contemporary Chinese thought*. World Scientific.
Wu, G. (2021). From the castrated subject to the human way: A Lacanian reinterpretation of ancient Chinese thought. *Psychoanalysis and History, 23*(3), 187–213.
Yang, S., & Rosenblatt, P. (2001). Shame in Korean families. *Journal of Comparative Family Studies, 32*(3), 361–375.
Zhu, Q. (2020). Ethics, society, and technology: A Confucian role ethics perspective. *Technology in Society, 63*, 101424.
Zhu, Q. (2023). Confucian ethics of technology. In G. Robson & J. Tsou (Eds.), *Technology ethics* (pp. 93–101). Routledge.
Zhuangzi. (n.d.). *Zhuangzi* (J. Legge, Trans.). Chinese Text Project. https://ctext.org/zhuangzi
Žižek, S. (1989). *The sublime object of ideology*. Verso.

Part IV
Problematizing the Western Narrative

Chapter 7
Ex Oriente Progressus? The Soviet Union, Between East and West, as the Light of the World

Christopher Coenen and Aleksandra Kazakova

Abstract If we reverse the question of what went wrong with 'the East' compared to 'the West' in terms of 'progress' and instead try to determine what went wrong with 'the West' in the face of past and impending disasters, we must first discuss relevant meanings of the metaphors 'East' and 'West' and clarify what kind of progress we refer to. This can serve as a basis for analysing a bygone 'East', the Soviet Union, which, despite many terrible and sobering experiences, was for numerous people around the world, including in 'the West', for several decades an 'East' in which much—or even, in a fundamental sense, everything—went right and truly human progress was made. The Russian Revolution and Soviet Russia were seen by many as a light of the world that shone from the 'East'—*ex oriente lux*—, and this also applied to the Soviet Union for much of the twentieth century, including in the 'Far East' and the 'global South'. However, the hopes associated with this new metaphorical 'East' changed from almost apocalyptic yearnings for imminent global liberation to expectations of Soviet support of the 'Third World', including East Asia, to catch up with 'the West' as effectively as the Soviet Union did in many respects. Cultural representations and practices of engineering and technology both played a key role in the rise of the Soviet new 'East' against the backdrop of an earlier history of East-West dichotomies and in the history of interactions between the Soviet Union on the one hand and the Asian East and the Global South on the other. The real history of comprehensive, i.e. technical-scientific and social progress must be reclaimed against Western ideology, and the critique of anti-'Eastern' and anti-'Southern' Eurocentrism should become a key element of materialist superstructure analysis in science, technology and engineering studies, with aspects of the history of religion and philosophy also being considered central. With regard to

C. Coenen (✉)
The Institute for Technology Assessment and Systems Analysis, Karlsruhe Institute of Technology, Karlsruhe, Germany
e-mail: christopher.coenen@kit.edu

A. Kazakova
University of Chinese Academy of Sciences, Beijing, China

L. Checketts (ed.), *What Went Wrong with the West: Carl Mitcham's Critique of Western Philosophy of Technology*, Philosophy of Engineering and Technology 52, https://doi.org/10.1007/978-3-032-22214-5_7

the latter in particular, Carl Mitcham's stimulating discussion of Chinese and Western technological and cultural imaginaries deserves further attention.

Keywords East-West dichotomy · The East · The West · Soviet Union · Russia · Germany · China · Geographical metaphors · Progress · Marx · Marxism · Superstructure · Eurocentrism · Carl Mitcham · Social democracy · Socialism

7.1 Introduction

In twenty-first century discourse on 'the West', the Western dominant narrative still—and in the 2020s increasingly—equates the West with progressiveness. Unlike in the previous two centuries, however, the concept of progress central to the discourse is now less about the state of technoscientific progress and more about cultural and political aspects such as individualism and liberal democracy. This is not surprising given the technoscientific progress in such non-Western countries as China. However, it is noteworthy because it conceptually separates scientific and technological supremacy from cultural supremacy, whereas previously the coupling of claims to both supremacies, with a focus on the former, was at the centre of Western ideological production (Adas, 1989).

If we want to discuss what went wrong with 'the West' in terms of past catastrophes such as the two World Wars, the impending global ecological catastrophes resulting from the Western model of capitalist development or the current renaissance of fascism, we are well advised to first discuss a number of conceptual and historical aspects concerning notions of 'the West' and 'the East'. Of course, we alternatively could take at least 'the West' as a given unit, defining it, for example, geopolitically as the NATO member states and their close allies. With such an approach, however, we would run the risk of overlooking highly relevant aspects of our question, as we would ignore the thorny issue of changing normative charges and political uses of geographical metaphors. In the following, we therefore first discuss relevant meanings of the geographical metaphors 'East' and 'West' when used as cultural and political terms, and where these terms originate.

As Antonio Gramsci noted, the ideological dichotomies of 'East versus West' are "objectively real"—and he himself puts this in double quotes—even if "analysis shows that they are only a conventional, i.e. 'historical-cultural' construction" (Gramsci, 1977, 874). He adds in brackets that the term 'conventional' here refers to 'historical' facts that are products of civilisational development and "not just rationalistically arbitrary or individually contrived constructions" (ibid.). Since historical content has been linked to the geographical terms, 'East' and 'West' now denote specific relations between different cultural complexes. Even if these references are often geographically absurd, they are real, "they correspond to real facts, they allow one to…'foresee' the future, to objectivise reality, to understand the objectivity of the external world" (ibid.). And he adds that "it seems that one cannot understand the philosophy of praxis, its position in comparison with idealism and with mechanical materialism, the importance and significance of the doctrine of superstructures" (Gramsci, 1977, 874f.) without having understood this relationship.

But do the 'East' and 'West' metaphors really refer to historical facts? We would argue that these (often vaguely) geographical metaphors, though used by some of the brightest minds since the nineteenth century, only point to facts when they express yearnings for a better world and human progress. However, such longings are currently marginalised to a large extent. Due to alienation and the weakness of emancipatory class struggles, immaterial culture and the entire superstructure are still tainted with too many harmful ideological deceptions to serve as a means for the self-liberation of the human race. In recent decades, especially since the 1990s and rapidly increasing in the 2020s, irrationalism has been growing rapidly, especially in the West (Coenen, forthcoming), and East-West dichotomies have a large share in that. In this ideological production, human and civilisational progress is often reduced to technoscientific advances that are or will be very unevenly distributed within humanity. Cultural and social progress appear negligible or of secondary importance at best, and large segments of humanity are viewed as useless obstacles to eternal, techno-capitalist progress.

After the end of the Soviet Union, elites in the West engaged in the construction of new enemies, focusing on Islamists or all Muslims. Recently, 'Asian hordes' have been added to Western media discourse again, in the form of soldiers of the Russian army. Except for the Covid pandemic, anti-Chinese resentments are not a central element in this inner-Western public discourse, in which the immediate Others are Africans, Arabs, Muslims and, in the case of the US, Latin Americans. In it, Western oligarchs, political leaders and ideologues are making ever more liberal use of the ideological armoury of colonialism, European and Western imperialism, nationalism, historical fascism, anticommunism, religious bigotry and even such cultural traditions as transhumanist and Tolkienian imaginaries—all in search for legimitisation of what is not justifiable anymore: Western global dominance. In this sad spectacle, in which all the traditional great powers with a dominant Christian tradition plunder the ideological poison vaults of the nineteenth and twentieth centuries, the often hateful, demonising use of 'East'-'West' metaphors is widespread.

However, the incoherence and demagogic nature of this ideological discourse should not prevent us from trying to analyse the 'real-life' significance of the metaphors, if only because they play a crucial role in the ongoing pro-capitalist global propaganda and information warfare. On a deeper level, these metaphors are not just 'opium for the people'. They often express yearnings for a good life, a better world or even salvation, but they are regularly used for geopolitical power games, repression and propaganda. In the ideological realm, there is a land of the living and a land of (un)dead ideas, and the bridge is scholarly practice—insofar as it is based not only on being an observer but also on being engaged with the struggles that began after the end of the Eastern bloc. Carl Mitcham's writings and scholarly practice of building bridges between continents are a prime example of this.

We were both born in states (the Soviet Union and West Germany respectively) whose societies, and even more so their predecessor states, were historically deeply shaped by references to Christianity and the Roman Empire. They are therefore, at least from a Chinese perspective, Occidental—to avoid here the term 'West', the current use of which was largely shaped by the Cold War. At the same time, the societies in question, the tsarist Russian, the Soviet, the modern German (1871 to today) and the post-Soviet Russian ones, have defined themselves to varying degrees through historical demarcations against 'the West', or still do. And they were and are often regarded in 'the West' itself as non-Western or not really Western—or in the case of Russia and the Soviet Union as Eastern or even as the main representatives of 'the East'. And crucially in this context, German and Russian discourses and 'Western' images of Germany and Russia have played key roles in creating the modern notion of 'the West'. As GoGwilt (1995) described it, Europe's both arrogant and deeply insecure imaginary political self-isolation in the era of Western imperialism brought with it a concept of 'the West' that oscillated between megalomania and paranoia and was built on a narrative of European history that defined the West in opposition to various 'Easts'.

To come back to Gramsci, we would argue that, beyond the sad and abhorrent fact that they often serve as a justification for post-socialist and post-colonial manslaughter, the politico-cultural notions of 'East' and 'West' are 'objectively real' only where they refer to actual human progress—be it cultural, technoscientific or political—whereby this concept of progress implies that all such advances have the potential to help overcome the capitalist world (dis)order. One of our aims here is to show that reactionary 'East versus West' metaphors stand in the way of such human progress.

To this end, we would like to recall that there was once another 'East' in global discourse, which was only considered to be 'catching up' with the West in terms of economic and technoscientific progress, while it was seen as more progressive in important cultural, social and political aspects such as social and gender equality or enlightened thinking. We refer here to the Soviet Union and more broadly the 'Second World' of socialist states, or 'East bloc', which most strongly represented a progressive 'East', also in terms of science, technology and engineering, immediately after what is known in the West as the 'Sputnik crisis' (and in West Germany as 'Sputnik shock'). It was back then, in 1957, that Mao Zedong's use of the metaphor of the 'the East wind prevailing over the West wind' (Li, 2019) became widely known outside China, after he had used this metaphor from classical Chinese literature during a Moscow visit to express his view that, after the launch of Sputnik, the socialist camp not only had an overwhelming advantage over the Western imperialist countries in terms of popular enthusiasm and numbers of people, but also acquired one in science and technology. The 'East' that Mao here referred to was the global socialist camp, including in a prominent position the Soviet Union, the factual successor state of one Christian European great power, the Russian Empire. The new, Soviet 'East', besides being seen also in the West by many as the homeland of the global proletariat, also established new relations with the 'Far East', i.e. East Asia, and with the global South. In this context, we refer to the Soviet Union with the metaphor of the 'Northern Prometheus' alluding to the Titan's role as a rebel against the Olympian gods, who took fire from them and gave it to the human race so that it could develop technology, increase its knowledge, and advance civilisation. We call it a 'Northern' Prometheus, not only to recall the old designation of the Russian Empire as a Northern power, but also to emphasise the position of the Soviet Union between 'the West' and the Asian East and its special role as a Northern country in the global South.

7.2 The Rise of a New East Against the Background of Earlier East-West Dichotomies

In what follows, we argue that for many, including in the West, the new, Soviet East embodied comprehensive human progress which made 'the West' appear backward, reactionary, and declining. This was particularly true during the early, "utopian" phase of the Russian Revolution and Soviet Russia. Later, despite of Stalinism, this hope by no means disappeared completely. However, its focus shifted, particularly as the Soviet Union emerged as one of the two superpowers. It was now increasingly less about the hope for a rapid global revolution anymore, but—given the Soviet Union's impressive industrialisation—about the expectation of primarily technoscientific and socio-economic (and only secondarily cultural and political) progress, of catching up with 'the West' within the national borders of an independent state not subject to Western imperialist control. Only at first glance, paradoxically, can the history of this new East be thus understood as a central element in a process of 'catching up with the West' or even Westernisation (in the sense of modernisation according to core features of the Western model) of large parts of the non-Western world.

First, however, we sketch the ideological and historical background against which the new 'East' rose. Historically, the modern concept of 'the West' emerges in Europe around the 1830s. Before, other politico-cultural notions of the 'West' or 'Occident' had existed, most prominently in the idea of a westward progression of both learning (*scientia*) and power, first developed, with creative reference to biblical and early Christian sources, by Otto von Freising. Through George Berkeley ("Westward the course of empire takes its way") and protestant clerics it strongly influenced US national ideology, such as the ideologeme of 'Manifest Destiny' (Schulte Nordholt, 1995). This idea is closely related to the concept of the *translatio imperii*, which is crucial in Western political thought and philosophy of history because it lends credibility to the assumption that Rome never ceased to exist but is continued in medieval polities and modern states up to the United States. A key assumption of Christian tradition is that Rome was necessarily the last empire before the overall triumph of Christ. From the Carolingians to the Russian tsars, the Roman Empire was invoked as the empire in whose tradition every empire must be—and also the two most important bourgeois republics, France and the United States, reveled in Roman nostalgia from the beginning of their existence.

In the Russian Empire before the early nineteenth century, non-geographical references to 'East' and 'West' were at least usually only made in religious contexts, either using the 'West' as a metaphor for the end of life or the Last Judgment, or by referring to the difference between Eastern Orthodox and Western Christianity (for this and the following, see Sdvižkov, 2015)—and Russia was seen, in and outside Russia, as a 'Northern' country, in a camp opposite to the 'civilised' South (of Europe). In discourse influenced by Enlightenment thinking, this could entail the hope that the sun of the Enlightenment coming from the South would eventually also enlighten the North, seen in line with old Greco-Roman ideas as barbaric territory.

However, emerging Romanticism changed that imaginary. Also parts of Germany, in particular those that had never been part of the Roman Empire, were widely perceived as 'Northern'. This was especially true for Prussia whose rise also increased cultural interest in it in other parts of Europe. The "inner-European 'North-South' divide" (Sdvižkov, 2015, p. 98) was further consolidated ideologically under the influence of the French Revolution and the Napoleonic Wars, in particular through a conservative Prusso-Russian 'Nordic solidarity' against France. Against the dichotomy of a fundamentally Roman civilisation versus a barbaric North, German and Russian authors set the ideologeme of a culturally and spiritually superior North attacked by barbarically acting Southern fanatics. Moreover, Germans (and again especially Prussians) emphasised a supposed special affinity of German with ancient Greco-Roman culture. This culturalist argument also served to stabilise a self-image in which German 'culture' was opposed to Western 'civilisation' and its 'materialism', with the political implication of equidistance from backward Oriental despotism and fanatical Western, particularly French, republican and democratic radicalism.

Soon after, "the geographical 'South-North' axis was largely displaced on European mental maps by one with a 'West-East' alignment" (Sdvižkov, 2015, p. 99). In the 1820s, Georg Wilhelm Friedrich Hegel revived the old idea that decisive intellectual and political progress was always made in demarcation from the East, stating that world history travels from East to West and that Europe is the absolute end of history, just as Asia was the beginning. Under his influence and against the backdrop of rising nationalism and widespread disillusionment with the universalist optimism of the early Enlightenment, Russian and German intellectual life developed into a hotbed of reflection on various East-West relations, running parallel to the soon-to-begin, more inward-looking debates in Western Europe about 'the West' or 'Occident', in which Auguste Comte played a significant role (Varouxakis, 2025). And in the decades that followed, well into the twentieth century, German and Russian discourse was strongly influenced by the respective perceptions of cultural affiliation with the East and the West, or relations with the West in both countries.

Inspired by German philosophy and Romanticism, Slavophiles and Westernisers in the Russian Empire both suggested often paradoxical positions on technology and culture and the links between technological, cultural and political progress. They helped construe the mutual 'otherness' of Russia and the West. In the 1830s, Pyotr Chaadayev's highly skeptical account of the Russian ability to catch-up with the Western parts of Europe initiated the debate. He portrayed Russia as being neither of the West nor of the East, as a country without a past and without a future, isolated from the rest of humanity and lacking any development of its own, any real progress. His polemic was based on a non-universalist vision of knowledge (living tradition, vast experience and intellectual skills), according to which the political and scientific developments rooted in the Latin-Catholic tradition were irreproducible in the Russian context. In 1857, the Westerniser Alexander Herzen warned against the technologisation of despotism without social modernisation, that if progress would only be realised by the government, it would be an unprecedented violence armed with science: "[s]omething like Genghis Khan with telegraphs, steamships, and railways, with Carnot and Monge on the staff, his soldiers armed with Minié rifles and Congreve rockets, and led by Batu Khan" (Masaryk, 1919, 419). Just as much as 'the West' was a vaguely generalised path of quite diverse scenarios of modernisation (British, French or German), the Westernisers' thought proliferated in quite diverse directions, as long as the common goal of abolishment of serfdom was achieved—with the utopian socialists, critical of contemporary Western capitalism, on the one hand, and the liberal reformists on the other. Early Slavophile critique of 'European rationalism' (under which capitalism, industrialisation and scientism were seen as inseparable) was partly opposed to, partly in line with Western irrationalist, religious, or aesthetic reactions to progressivism of that time. These different dimensions of anti-industrialism were expressed, for example, by the Slavophile Ivan Kireyevsky (1806–1856) in the final year of his life when he bemoaned that industry rules everything in the world, without faith and poetry, and that men bow down before it and erect temples to it as the real deity in which people sincerely believe and to which they submit (Christoff, 1972). Late Slavophiles around 1900 had to adjust this critique to the interests of the national bourgeoisie and to make their philosophy useful for the imperial bureaucracy. They did so by promoting protectionist policies against free trade, also with an anti-German thrust, and by developing pro-rural utopian visions, some of them pro-technology and kind of left-wing, in which the state had to find techno-moral solutions for the problems of class struggle, the centralisation of capital and the crisis of Russian identity (Suslov, 2012). But views as the ones expressed by Kireyevsky for a long time allowed for a Russian demarcation against a soulless and materialist Western industrialism.

With the revolutionary struggles in the West—including in the German states—in the mid of the nineteenth century, Russian perceptions of Germany also tended to see the latter as an example of Western individualism and political radicalism. While Russian liberal Westernisers sometimes sympathised with the West, juxtaposing it with an Asian, backward East, Slavophiles tended to perceive Russia and the non-Slavic parts of Europe as "fundamentally different civilisations", imagining "Russia as a truly Christian—that is, European—civilization, which differed fundamentally from both the Asiatic 'East' and the modern 'West'" (Sdvižkov, 2015, pp. 99f.) and condemning Western individualism and political divisiveness. At the same time, German revolutionaries, including Engels and Marx, took anti-Slavic stances due to the Russian invasion in Hungary and disappointment with Slavs in Austria-Hungary.

A key event in Russian self-definitions towards 'the West' and in particular the German world was the Crimean War (Sdvižkov, 2015): Prussia remained neutral, but in defiance of the old Nordic solidarity, and Austria-Hungary, while not directly attacking Russia, used the opportunity for territorial gains. In the course and aftermath of the Crimean War, significant parts of the Russian elites now defined Russia as (a European) 'East' in confrontation with 'the West". In the following years, due to geopolitical dynamics, rising culturalism, nationalism, and the geographical proximity of the German and Slavic worlds, anti-German ideology in Russia grew stronger, often with anachronistic reference to the medieval Eastern expansion of German-speaking settlers and polities. In the struggle for 'Eastern Europe' or, alternatively, between Europe and Russia, the 'Germanic tribe', embodied foremost by Austria and Prussia, became Russia's main opponent, and a "key role in consolidating feelings of enmity was played by the slogan *Drang nach Osten*, referring to Germany's 'thrust to the East'" that "initially emerged in Polish discourse towards the end of the 1840s and entered Czech and Russian discourses during the 1860s and 1870s" and "inverted the widespread Western notion of an expansionist Russia" as the greatest threat to the Occident making raids on the West (Sdvižkov, 2015, p. 102). However, in contrast to the old identification with 'the North', Russian self-conceptions of belonging to 'the East' were ambiguous and often combined a self-understanding as European with a demarcation from the non-Christian Asian East. The poet and philosopher Vladimir Solovyov—whose often violent anti-Asian (anti-Chinese and anti-Japanese) rhetoric can be deemed an intellectual prelude to the Russo-Japanese War (1904–1905)—wrote a poem in 1890, entitled, by using an old Christian image, *Ex Oriente Lux*, whose combinations of metaphors and historical references are very telling in our context:

"Light comes from the East, strength comes from the East! / And, ready to reign, / The king of Persia drove his herds of slaves / To Thermopylae. / But the heavenly gift of Prometheus / Wasn't given to Hellas in vain. / Paling, the crowds of slaves retreated / Before a handful of valiant men. / And who has traveled the glorious path / To the Indus and the Ganges? / The Macedonian phalanx, / The sovereign eagle of Rome. / And by the strength of reason and law—/ Truths undenied by any man—/ An empire of the West arose, / And Rome unified the world. / What more was needed? / Why was the whole world again in blood? / The soul of the universe craved / The spirit of faith and love! / And the prophetic word was not false, / And a light from the East shone, / Heralding and promising / What had been impossible. / And, spilling wide, / Full of portent and strength, / That light from the East / Made peace between East and West. / O Rus'! In lofty premonition / You ponder a proud idea; / Which East do you want to be: /The East of Xerxes or of Christ?"

In this vision, Russia has the chance to play the role of a truly Christian heir to the Roman Empire, making peace between East and West in Europe while at the same time staying strong against a barbaric and despotic Asian East.

The poem was created in a new phase of German-Russian ideological 'entanglements' (Sdvižkov, 2015), in which early conservative hopes for a German-Russian alliance after the Prussia-led founding of the first German national state in 1871 had been replaced by an ideologically even more aggressive enmity. While a further "differentiation of the concept of the West according to political camps" took place that "also created a new basis for mutual conceptual transfers" (Sdvižkov, 2015, p. 102) across borders, the so-called Second German Empire, deemed by many inside and outside Germany as situated between 'East' and 'West', was increasingly seen in Russia, together with Austria, as an enemy much worse than the old 'West'. While remnants of 'Nordic solidarity' remained in some corners and partly even intensified (particularly in the cultural sphere), both states increasingly became outright rivals against the backdrop of Russia's rapprochement with the Western powers, which were alarmed by Germany's rise to power. Conservatives and liberals in Russia could agree on an anti-German position, and the same is true to some extent for Germany in terms of anti-Russian resentment. And in both countries, liberals tended to glorify or at least highly value 'the West', often labelling the other country as insufficiently Westernised bulwark against cultural and political progress.

Against this ideological background, a new East entered the world stage in the year 1905 with the start of the first Russian Revolution. At the same time, and perhaps as a necessary precondition, a major European power, namely Russia, lost a major war against a non-Western state, namely Japan, a fact that shocked many in the traditionally Christian countries that had become strongly racist in the course of the nineteenth century. The rise of a modern, technologically advanced Asian East thus stood at the cradle of the new revolutionary European East.

7.2.1 The Rise of the New East

Thanks to the rise of social democracy, Marxism, class analysis, and anti-imperialist and other emancipatory struggles, a framework for understanding history and the world had emerged that simultaneously represented a radical break with the old Christian tradition and opened up a perspective for also overcoming the more recently developed bourgeois nationalism. In this process, the East-West metaphors were ideologically recharged.

This emergence of the new East was clearly articulated by the important social democratic theorist Karl Kautsky even before 1905. In an article "The Slavs and the Revolution", first published in Russian in the journal *Iskra* when Lenin was still one of its editors (Kautsky, 1902), he historically contextualised the early anti-Slavic resentments of Engels and Marx, arguing that due to the course of events in 1848 and 1849 it was not surprising that the German revolutionaries were seized by such a fierce hatred for the Slavs that they began to look upon them as degenerates among the nations, as a nation of slaves, to be stepped over by the revolution. Such resentments were in line with the traditional thinking that the East was a stagnant or even reactionary force in world history. Applying Marxist analysis, Kautsky rejected the idea that a mysterious predisposition to slavery could explain the anti-revolutionary behavior of Slavs. With capitalism now having spread to large parts of the East, the Slavs possessed revolutionary classes too. And not only a great cultural struggle was now being waged in the Slavic world for the liberation of the proletariat, and thus of all humankind, but the centre of gravity of socialist revolutionary thought and action further shifted from the West to the East: having been in France until 1871 (tragically culminating with the Paris Commune), afterwards in Germany with the impressive rise of social democracy, and now moving further to East. Russia, having absorbed so much revolutionary initiative from the West, may now perhaps be ready to serve as a source of (new) revolutionary energy for Western Europe too. Kautsky emphasised that Russia had long since ceased to be a mere stronghold of reaction and that now Western Europe was developing into one, its bourgeoisie pumping capital into an autocratic and backward regime in Russia.

When some of Kautsky's expectations materialised three years later, Rosa Luxemburg said in a speech covered by a newspaper and protocolled by a spy of the German Empire (Luxemburg, 2005) that political freedom is already emerging in Russia and that the revolution, although it will probably take quite a long time, has already taught the social democrats everywhere the direction to go. And she used the metaphor that the Russian revolution had already cast its fiery light on other countries. The new, Soviet East—Trotsky and others had created the first Soviet in Petersburg also in the year 1905—had emerged in Russian as well in German discourse, and beyond.

However, Kautsky would not have been Kautsky if he had not also clearly articulated the fatal, reactionary East-West dichotomisation early on that later served as the ideological justification for the First World War and thus the politico-moral bankruptcy of large parts of Western and especially German social democracy. In 1907, he wrote that the triumphal march of democracy has proceeded very slowly from West to East, and as often as it had achieved significant progress in the West and had come to dominate a state, it was threatened by Eastern despots. Kautsky argued that compared to the French Third Republic, the Prussian-German forces in 1871 were a reactionary power whose policy of conquest it was the patriotic duty of every democrat in France to reject, but that compared with Russian absolutism, the Second German Empire represented an enormous advance, writing that "just as it was the duty of every French socialist to develop his patriotism to the utmost in the war of the Republic against the Empire, so the same duty would have been incumbent on German socialists if it had come to a war between Germany and Russia" (Kautsky, 1907, 14; *own translation*). He added that Russia itself has ceased to be a danger to the "freedom of the West" since the outbreak of the revolution.

In the declaration of the German Social Democratic Party in the Reichstag (i.e. the German parliament of the Second German Empire) after the start of the war in 1914, the key argument for the support of the war is that "[m]uch, if not everything, is at stake for our people and its free (*freiheitliche*) future in the event of a victory for Russian despotism, which has stained itself with the blood of the best of its own people" and that "[w]e must…safeguard the culture and independence of our own country [by] doing what we have always emphasised: we will not abandon our own country in its hour of danger" (Deutscher Reichstag, 1914, 8f.; *own translation*).

Although the other parts of the declaration adhere to Marxist class analysis, the critique of imperialism and proletarian anti-war convictions, this resumption of an anti-Eastern rhetoric that equates the East, epitomised by the Russian Empire, with despotism and mentions the Russian revolutionaries only as victims, served—together with (moderate) nationalism—to dismiss the new framework for understanding history and the world at a moment when it would have been most useful in preventing a global catastrophe. This provided an ideological justification of a policy of support for a full-blown war between the major imperialist powers in Europe, in which, as Walter Benjamin wrote later, an "immense wooing of the cosmos was enacted for the first time on a planetary scale, that is, in the spirit of technology" and "technology betrayed man", turning "the bridal bed into a blood bath" due to "the lust for profit of the ruling class" (Benjamin, 1928/1978, 93; cf. Coenen & Kazakova, 2021).

Many German bourgeois intellectuals and scientists indulged in militarist and nationalist euphoria, hoped like the eminent German writer Thomas Mann for an end of decadent 'civilisation' through a cleansing war experience, and set German 'culture' against 'civilisation' as well as against the English 'merchant's soul', French superficiality and radical democracy, and Russian despotism. Much of this propaganda soon focused on the 'West' because after the atrocities of the German troops in Belgium the German self-understanding as a 'nation of culture' and its image as a global beacon of science and the arts had come under massive pressure and so educated German felt more compelled to defend themselves against the West than to polemicise against an East that was widely seen as barbaric, uncultured and uncivilised anyway. However, in the earliest major expression of the so-called 'ideas of 1914', the *Manifesto of the Ninety-Three*, signed by artists, scholars and scientists, the polemic against the Western powers was supported by anti-Eastern resentments, coupled with racism: "In the East, however, the blood of women and children slaughtered by Russian hordes soaks the earth, and in the West dumdum bullets tear the chests of our warriors. Those who ally themselves with Russians and Serbs and offer the world the disgraceful spectacle of setting Mongols and Negroes against the white race have the least right to pretend to be the defenders of European civilisation" (*own translation*).

This propaganda was a foretaste of what was to follow in the Second World War during the German War of Annihilation in the East and then during the Soviet counter-attack. In the years 1941–1945, Slavs were often characterised as racially inferior or even as subhumans, the people of the Soviet Union and in particular their communist elite as *verjudet* ('jewised', i.e. with a strong Jewish element) and thus as subhuman and as arch-enemies, and the non-European Soviets as Asian hordes against whom 'Fortress Europe' had to be defended under German leadership from 1943 onwards.

In 1914, the situation was still different. In his speech to the parliament after the start of the First World War, Kaiser Wilhelm II regretfully recalled the old Nordic solidarity—the past brotherhoods in arms with the Russian Empire ("It was with a heavy heart that I had to mobilise My army against a neighbour with whom it had fought together on so many battlefields.")—, castigating only Serbian nationalism. And many members of the bourgeois and aristocratic elites ideologically positioned Germany as a power between East and West defending Europe and its "highest culture" against both. On the other hand, the anti-Russian propaganda was not only a liberal bourgeois endeavour, but also one in the German social democracy, which soon split in a way similar to the Russian social democracy (Mensheviks versus Bolsheviks).

The love for the new East was fervently expressed by left-wing social democrats and other radical anti-capitalists who opposed the war and of which some founded the Communist Party of Germany after the war. During the war they cooperated first in left-wing groups within the Social Democratic Party of Germany and then in the newly founded Independent Social Democratic Party of Germany. The most famous person engaged in these circles is Rosa Luxemburg who was imprisoned in 1915 and wrote in this year the so-called *Junius Pamphlet* (Luxemburg, 2000). The actual title of the work, published in 1916, is *Die Krise der Sozialdemokratie* (The Crisis of Social Democracy). In it, she fiercely criticised German Social Democratic anti-Russian war propaganda, analysing the new East in Kautsky's tradition but in a much more detailed, more differentiated and at the same time even more glorifying way. In her view, the declaration of the Reichstag Social Democrats differed from the government declaration only in its focus on 'Russian despotism' and abused revolutionary traditions to democratically ennoble the war.

However, in addition to the old progressive anti-tsarist tradition, the right-wing social democrats were also able to build on later statements and some concepts of Marx and Engels. In 1867, for example, Marx had said in a speech that the German proletariat "will be forced by its geographical situation to declare war on oriental barbarism, for it is from Asia that the whole reaction against the West has come" (cit. in Shanin, 1983, 143). Here, it is important to notice that quotes and concepts were often either misinterpreted or torn out of their historical context.

7.2.2 Marxian Views on Different 'Easts' and Social Democratic Anti-Eastern Propaganda in 2014

In our context, it is in fact crucial to understand what role the East—both Russian and Asian—played in Marx's thinking and to what extent it was Eurocentric. We cannot discuss this at length here, it must suffice to briefly justify why we believe that some of his anti-Eurocentric critics do not give full justice to several aspects of his thinking and ignore how it changed over time (cf. Lindner, 2010).

Together with his changing views on Russia, most important here are his thoughts about types of pre-capitalist societies, their periodisation, the 'Asiatic mode of production' including 'Oriental despotism', 'barbarism' and 'civilisation', which he often developed not only with reference to Russia but also India and China. A key aspect of Marx's evolving thinking about these topics is the fact that from the 1870s on, he increasingly focused on specificities of Eastern societies (in China, India and Russia) that could not easily be subsumed under a global unilinear developmental scheme from primitive communism via slaveholder societies and feudalism to capitalism. Another important aspect is that already before the 1870s, Marx had developed the notion of an Asiatic mode of production in which Oriental despotism was the political key element and self-sufficient agrarian communities were a presumedly timeless socio-economic structure. Already then, he was thus not only looking at the East with Western lens. However, some of his anti-Eurocentric critics argue that this conceptualisation is a prime example of his Eurocentrism because it assigned Asia a passive role in world history and deemed Asian societies as inherently non-progressive. In this view, Marx simply continues the Hegelian views of the East and the West in world history. Last but not least, some critics cite Marx's use of the terms 'barbarism' and 'civilisation' as further evidence for his Eurocentrism. Here, referring to quotes where they serve to characterise large societies at the periphery of Western-led capitalism, particularly in China and India.

In our context, the following aspects of Marxian thinking are most relevant: From 1870 until his death in the year 1883, he clearly overcame simplified views of Russian, Indian and also Chinese history and societies as well as a schematic Hegelian philosophy of history (cf. Lindner, 2010). Key for that was his keen interest in the question of village communities and agrarian communal land ownership, which also led him to learn Russian.

In earlier writings on India in 1853, he had in fact portrayed, in line with the *Communist Manifesto*, the Western capitalist work of destruction of the Eastern societies as a necessary step in human social progress, characterising England as an "unconscious tool of history" in bringing about "a fundamental revolution in the social state of Asia", without which humanity could not "fulfil its destiny" (Marx, 2008, 218f.). But even then, he left no doubt that Western despotism in India was worse and more barbaric than the traditional Oriental one. Here it is important to notice that Marx used the notions 'barbarism' and 'civilisation' by no means only, and not even mainly, to juxtapose Western civilisation with barbarism. In his view, later taken up by Luxemburg and many others, barbarism is always an element within Western capitalism: "The profound hypocrisy and inherent barbarism of bourgeois civilization lies unveiled before our eyes, turning from its home, where it assumes respectable forms, to the colonies, where it goes naked" (Marx, 2008, 224). And it was the case of Ireland (cf. Lindner, 2010) that led Marx to fundamentally question his earlier beliefs about a necessarily progressive role of British imperialism and, as another break with Eurocentrism, also made him first think of the option of globally significant revolutionary action starting in a colonised country.

When it comes to China, anti-Eurocentric critics of Marx also tend to oversimplify his thinking. Of course, his journalistic descriptions of the Taipings can be deemed strong evidence for Eurocentric and even racist elements of his mindset (Leung, 2013). In them, he entirely ignores the revolutionary agency of the Chinese insurgents, indulges in exoticism, and does not at all provide a materialist analyse of this civil war, explaining it solely with Western imperialist influence and resorting to culturalist musings in line with his concepts of an Asiatic mode of production and its eternal despotic character. On the other hand, in other writings about China, he popularised the notion of the 'Opium Wars' and strongly emphasised that Western imperialism in China was not at all a force of human progress (see, e.g., Marx, 2008, 23). In general, his hatred of European bourgeois and aristocrats was much stronger than his contempt for certain non-Europeans. Referring to the Chinese Emperor, he wrote, for example, that while "the semi-barbarian stood on the principle of morality, the civilized opposed to him the principle of self. That a giant empire, containing almost one-third of the human race, vegetating in the teeth of time, insulated by the forced exclusion of general intercourse, and thus contriving to dupe itself with delusions of Celestial perfection—that such an empire should at last be overtaken by fate on [the] occasion of a deadly duel, in which the representative of the antiquated world appears prompted by ethical motives, while the representative of overwhelming modern society fights for the privilege of buying in the cheapest and selling in the dearest markets—this, indeed, is a sort of tragical couplet stranger than any poet would ever have dared to fancy" (Marx, 2008, 27). And this rejection of the notion of British imperialism being a positively civilising force in the East did not only refer to moral values but also to socio-economic aspects. Marx variously emphasised that Western imperialism prevented the further development of capitalist modernisation in the colonised or half-colonised ancient civilisations of China and India.

While it should be obvious by now that Marx not even in his early writings simply reproduced Hegelian or other European culturalist East-West dichotomies and that in his writings on China and India he did not operate with the notion of eternally passive Eastern societies, the differentiated character of his understanding of East-West relationships in terms of revolutionary action and potentials becomes entirely clear in his reflections on Russia from the 1870s on.

In his geopolitical writings, Marx continued and even intensified his anti-Russian statements, but now not primarily to defend Western European progressive and revolutionary developments against the old bulwark of reaction, but to attack the tsarist regime as the weakest point of the global old order. In 1877, Marx supported Turkey against Russia and predicted revolution in Russia as the eventual outcome of the war. In line with the hope for a Russian revolution, Marx wrote in a private letter in September that "[t]his time the revolution is starting in the East, where the hitherto unshattered bastion and reserve army of the counterrevolution are situated" (cited in Shanin, 1983, 56). In his pamphlet whose title reads in English as "The Eastern Question or Will Europe Turn Cossack?" (Liebknecht, 1878), the important German social democratic leader Wilhelm Liebknecht agitated against Russia with a strongly dichotomous narrative of a confrontation between a despotic East (tsarist Russia) and a democratic and progressive West (the social democracy of the Western countries) and included quotes from a recent private correspondence between Marx and him on the Russian-Turkish war. These quotes also clearly show, however, that Marx's support for Turkey was only motivated by the hope that a Russian defeat would lead to revolution in Russia and that this will then trigger revolutions in Germany and the entire West.

Around this time (i.e. not earlier than 1877), in an unpublished letter to the Russian journal *Otechestvennye Zapiski*, Marx also emphasised that his sketch of the genesis of capitalism in Western Europe should not be understood as a "historico-philosophical theory of the general course fatally imposed on all peoples, whatever the historical circumstances in which they find themselves placed, in order to arrive ultimately at this economic formation which assures the greatest expansion of the productive forces of social labour, as well as the most complete development of man" (cited in Shanin, 1983, 136)—no such supra-historical "master-key of a general historico-philosophical theory" (ibid.) exists. In order to save the Russian commune, Marx wrote, there must be a Russian revolution and this was "the finest chance ever offered by history to a people" to avoid "all the fateful vicissitudes of the capitalist regime" (cited in Shanin, 1983, 135).

Marx's thinking about the Russian society and its potentially world-revolutionary role culminated in the context of his correspondence with the revolutionary Vera Zassoulich in 1881. While in the end he responded only very briefly to the latter's letter to him, the drafts of his reply—in which he polemicises against pseudo-Marxist "Russian admirers of the capitalist system" and argues that "[p]recisely because it is contemporaneous with capitalist production, the rural commune may appropriate all its positive achievements without undergoing its...vicissitudes (cited in Shanin, 1983, 106)—show how Marx's thinking on the topics in question had developed in the 1870s.

In her letter, Zassoulich had asked him to explain whether he thinks that the rural commune in Russia could become the collectivist basis for a socialist development or if it is "historically necessary for every country in the world to pass through all the phases of capitalist production" and thus "Russian peasant's land" has first to "pass into the hands of the bourgeoisie", followed by many centuries of catching up of Russia with "the level of development already attained in Western Europe" (cited in Shanin, 1983, 98). In his reply, Marx apologised for its shortness referring to a recent illness and pointed out that in the *Capital* he had written that the genesis of capitalist production—the complete separation of the producer from the means of production, whereby the expropriation of the agricultural producer is the basis of the whole process—had only in England been accomplished in a radical manner and that all other Western European countries are following the same course. And he emphasised that the "'historical inevitability' of this course is therefore expressly restricted to the countries of Western Europe" (cited in Shanin, 1983, 124). In the "Western case", one form of private property was transformed into another form of private property. In the case of the Russian peasants, however, "their communal property would have to be transformed into private property" (ibid.). Marx thus concluded: "The analysis in Capital therefore provides no reasons either for or against the vitality of the Russian commune. But the special study I have made of it, including a search for original source material, has convinced me that the commune is the fulcrum for social regeneration in Russia" (ibid.).

In the drafts of his reply (cf. Shanin, 1983, 99–123), the differentiated character of the late Marx's understanding of 'the East' becomes even clearer. Russia was for him the only European country in which the agricultural commune had maintained itself on a national scale and at the same time not become, different to India, the prey of a conquering foreign power. Nor did Russia live in isolation from the modern world. The way to go would be to transform the rural communal bodies into peasant assemblies chosen by the communes themselves, an economic and administrative body serving their own interests. And he argues that in India the suppression of communal land ownership was nothing less than an act of English vandalism that drove the population backward rather than forward. In his view now, British imperialism was not at all a force of social progress in this regard.

In the preface to the second Russian edition (1882) of the *Communist Manifesto*, which anticipates the global situation in the 'short twentieth century', Marx and Engels pointed to the decisive roles now played by the United States and Russia, roles very different from the mid of the nineteenth century when both were just "pillars of the European order" (cited in Shanin, 1983, 138). In the preface, Russia is not anymore the strong bulwark of reaction but the main target of revolutionary action: "Russia forms the vanguard of revolutionary action in Europe" (ibid.). And they wrote: "Can the Russian [...] primitive communal ownership of the land pass directly into the higher, communist form of communal ownership? Or must it first go through the same process of dissolution which marks the West's historical development? Today there is only one possible answer. If the Russian revolution becomes the signal for proletarian revolution in the West, so that the two complement each other, then Russia's peasant communal land-ownership may serve as the point of departure for a communist development" (cited in Shanin, 1983, 139).

This expectation became a key hope for the global socialist and then communist movements in the following decades. While the influence of the rather schematic and geopolitical thinking of the likes of Kautsky and Wilhelm Liebknecht, in which a "class-blind" Western solidarity against Eastern despotism was central, continued to be influential in the social democratic movement, the hope that in the (Russian) East global revolution will start became another key element of the thinking of this movement. In a letter to Laura Lafargue in 1882, Marx wrote on his growing influence in Russia that "nowhere is my success more delightful; it gives me that satisfaction that I damage a power, which, besides England, is the true bulwark of the old society" (cited in Shanin, 1983, 71). The 'old society' is here the global capitalist (dis)order and the West is not seen as the force of progress anymore.

To sum up the mature Marx's position: As part of a world of emerging global capitalism, the societies of 'the East' are by no means passive recipients of fundamentally beneficial Western modernisation, their specificities allow for other paths to communism than the Western one, and they can potentially play major and even decisive roles in the global revolutionary process. There are path dependencies but there is no supra-historical determinism that every society must follow and there is no reason for Western cultural chauvinism. Barbarism is a key element of bourgeois civilisation, in particular in the colonised or semi-colonised world, where at the same the forces of global human progress are rapidly gaining strength.

When during the First World War, Luxemburg criticised the abuse of socialist revolutionary traditions to democratically ennoble the war, she was thus right. And her anger also hit the many social democratic newspapers that supported the war.

The newspapers quoted in the *Junius Pamphlet* (Luxemburg, 2000), spoke of Russian rule with the rod and bestiality against Germany's women and children, that German land must not become the prey of the Cossacks, because if Germany were defeated, it would not be an English governor or a French republican who would rule over Germany, but the Russian tsar, a ruthless and barbaric enemy, against whom everything that existed in terms of German culture and German freedom needs to be defended. It was argued that tsarism had been rejected by German social democracy since Marx and Engels as a bloody refuge of European reaction and that now there was an opportunity to settle under the German banners of war accounts with this terrible society. In this battle of culture against unculture, German proletarians would also fight like men. In addition to the duty of national self-preservation, social democratic workers would also know that the enemy in the East was the enemy of all progress and culture. The defeat of Russia would become the victory of freedom in Europe. The class-conscious workers would also obey their own convictions when they defend the ground on which they stand from the invasion of the East. 'Down with Russian blood-tsarism! Down with the stronghold of barbarism! Against Russian despotism and insidiousness!' should be the paroles. The whole of Western Europe would have a vital interest in eradicating the hideous, murderous tsarism and the capitalist classes of England and France only acted against that in order to have the profit opportunities that German capital had previously enjoyed in Russia. In the age of capitalism, there would certainly be class struggles even after the great war but they would take place in a freer state and far more limited to economic areas, and the treatment of social democrats as outcasts, as second-class citizens, as politically disenfranchised, would be impossible in the future when Russian tsardom, the mortal enemy of democracy and freedom, has disappeared. Through a war for culture against the demonised enemy in the East, the proletarians will become free, 'freedom' meaning here above all more recognition as citizens in the capitalist state and a progressive culture mainly associated with the West. "So it echoed", wrote Luxemburg, "from the German party press in a polyphonic chorus. For people, however, who consult not the rabble instincts and the large-calibre slogans of the nationalist inflammatory press calculated for the rabble, but political points of view, it must be clear that Russian tsarism could pursue the goal of annexing Germany as well as that of the moon. At the head of Russian politics are cunning scoundrels, but not madmen' (Luxemburg, 2000, 96).

7.3 The New East as Revolutionary Light of the World

Towards the end of the Great War, when the revolution had broken out in the Russian Empire and the German Reich wanted to profit from it, Luxemburg and her Spartacus comrades called on the proletariat in Germany not only to support the Russian comrades, but to take them as a model. In *Spartacus Letter* no. 4, published in April 1917, Russia is called a "source of moral light in the East", the last corner where revolutionary socialism still had a course. And in January 1918, *Spartacus Letter* no. 8 the German workers were called upon to carry the message of revolution and peace from the East to the West. In Luxemburg's view, the West(ern working classes) thus had to catch up with the (Russian) East in terms of political progress. While the German bourgeoisie and most national social democratic leaders had positioned themselves against a supposedly barbaric East, sometimes using Orientalism and racism, left-wing and other uncorrupted social democrats and later communists turned to the new East in the hope that revolution and thus global social and cultural progress would come from it. As the prominent German expressionist poet and communist activist Johannes R. Becher wrote: "The light grows in the East. / May the poet swell towards you. / Sink, night!".

It is often forgotten now, in particular due to the confrontation between the EU and Russia since the 2010s, how strongly first the October Revolution, then Soviet Russia and finally the Soviet Union were perceived of as such a light from the East, as a source of quasi-religious but at the same time real-world hope for masses all over the world, including in the West. After Luxemburg and her comrade Karl Liebknecht, the second leading light of the younger social democratic generation in Germany, had been killed by proto-fascist militia acting on behalf of the right-wing Social Democrats in government, the hope for a German socialist revolution was not yet gone. In fact, it was a strong motivation for the war efforts of the Bolsheviks who now ruled Soviet Russia and whose large majority including the most important leaders Lenin and Trotsky saw it as a prerequisite of a world revolution and the future existence of Soviet Russia.

The death knell for the hope of an imminent world revolution was the year 1920. In Spring, the last major attempt of an anti-capitalist revolution was crushed in the industrial West of Germany, the Rhine-Ruhr area, where workers in arms had helped save the Weimar Republic from a right-wing coup d'état (the *Kapp Putsch*) and then some of them tried to realise a proletarian revolution. After the defeat of the Red Ruhr Army, which included communist, left-wing social democratic, anarchist and other activists, White Terror was unleashed in the proletarian strongholds of western Germany. Even nurses of the Red Ruhr army were bayoneted by students. Proletarians who had just helped save a weak government led by right-wing Social Democrats (in coalition with bourgeois liberal and Catholic parties) were brutally suppressed, as in a prelude to the fascist violence of the following years. In summer, however, the ongoing Polish-Soviet War developed favourably for the Red Army. The second Comintern congress, convened in Moscow in July, was both an expression of the Soviet Russian self-confidence to be the vanguard of world revolution and the hope that the Red Army would march to the West and reach Germany. In August, it reached the suburbs of Warsaw but then lost against the Polish army in the Battle of Warsaw, in particular in Poland and the Catholic Church also known as the 'Miracle on the Vistula'. In our context, it is important to notice that many in the West did not perceive of this fight as one between Russia and Poland but as a decisive battle between the global bourgeoisie and the global proletariat, where the hope for liberation came from the East.

Alastair Bonnett (2004) argues that from 1920 on, there was a transition in the Soviet Union (1) from imagining the West to be the home of socialist revolution to seeing it as the home of bourgeois reaction, (2) from understanding Marxism-inspired revolution as a form of Westernisation to understanding it as antithetical to Western values, and (3) from the proletariat being connoted and idealised as Western and European to the proletariat being defined as Russian and non-Western. The West was identified with such social ills as racism. Bonnet also emphasises that the Russian Revolution was key in the rise of the concept of 'the West'. While long been in use in Russia, its spread in Western Europe accelerated from 1917 onwards, soon becoming inseparable from a political identity of anti-communism and pro-capitalism. This also included a racist component, as the idea that Russia was only superficially European—like in Napoleon's aphorism 'Scratch a Russian, find a Tatar'—was cultivated in anti-Soviet propaganda at least until the 1960s.

And in the West, from 1920 onwards, even among representatives of Western social democracy who were—often very strongly and sometimes in a very Eurocentric, Orientalist way—critical of the Bolsheviks, quite a few looked to the Soviet Union as a source of hope for a future victory of socialism (Schöler, 2021). At the same time, in China, in the wake of the May Fourth Movement, as well as in many other countries in the East and South, communist rule on the former territory of the Russian Empire was increasingly seen as a model for the development of their own societies, in confrontation with a capitalist West that, in the eyes of many, now embodied full-blown reaction and anti-progressiveness. In Asia, Africa, Latin America and elsewhere, catching up with the Soviet Union became the goal, not catching up with 'the West' (or only in terms of technological progress and prosperity). In this respect, the USA, whose breathtaking technoscientific and economic development was seen as a model even in the young Soviet Union itself, became the only serious global competitor to the latter.

To give just one example of how the Soviet Union was still (and sometimes even increasingly so) seen as a light shining from the East after the end of hopes for an imminent world revolution, we can point to the image of the socialist country among Western academics, particularly in Britain. Following the shock of the world war and further propelled by the rise of fascism, the Great Depression and other symptoms of capitalist crisis, proponents of technoscience in the West argued for the modernisation of the science system, for a more wide-scale use of science for societally beneficial engineering projects, and for an overcoming of irrational anti-science ideologies, looking to the (Soviet) East almost like to a Promised Land. While some of these scientists, such as Desmond Bernal, had already been active as communists before, this tendency gained even more momentum after the participation of a strong Soviet delegation, headed by Nikolay Bukharin, in the Second International Congress of the History of Science and Technology in London in 1931 (Ienna & Rispoli, 2021). It is noteworthy that Joseph Needham, a key figure in Western and to some extent Chinese scholarly work on Chinese history particularly of science and engineering, was involved in these developments. The Social Relations of Science movement that emerged in this context was, however, not an exclusively socialist or communist phenomenon but also involved scientists, science managers and science communicators who were not anti-capitalist. Some of them almost glorified—and sometimes even more than communist and socialist scientists—the Soviet scientific system and the application of science in the socialist state to technology and society.

Already here, foreboding the image of the Soviet Union from the 1950s on, the new (Soviet) East was now not only seen as the homeland of the global proletariat and potential starting point of world revolution but as a force of reason, rational planning and modernisation that favourably compared to an old (European) West, seen as chaotic, irrational and backward. In the words of an Argentinian visitor to the Soviet Union in 1935: "The traveler arriving in Russia after having passed through the Jesuit Spain of Gil Robles, the France of the decree-laws, the vast concentration camp of Germany, the tortured and martyred Poland, is suddenly seized—as if the landscape were suddenly changed—with the impression of living in another world, of breathing in another atmosphere, of treading on another land. It might be said, indeed, that he had escaped from his time, and that, like in a Wellsian fantasy, he had been given to anticipate his time, to approach the future, to stand upon the centuries to come. He has left behind him a society that is bleeding to death in misery and opprobrium, a society in which the unemployed number in the millions, in which intelligence is muted and culture humiliated, in which science is stopped except in the service of war, in which the very 'human rights' that the bourgeoisie promised for more than a century ago are denied and scorned, and in which the consciousness of its own ignominy has reached such a point that it has not hesitated to confess through the mouth of a former president of the French Council of Ministers that at the present time it is indispensable to 'put Prometheus back in chains'" (Ponce, 2007; *own translation*). Against this background of an increasingly reactionary and willingly backward West, the Soviet East is then praised by the author as a shining example of what humanity, freed from capitalism, is capable of in terms of science, engineering and culture.

For significant parts of the world population, including in the West, the Soviet new East embodied dynamism, social equality including women rights, modernity, the fight for global justice, decolonialisation, societally beneficial technoscientific progress, and thus the future. It was not until the 1980s that this East lost most of its appeal and 'the West' became the target of hopes beyond its geographical confines, even in parts of the world where many were fully aware that the concept of 'the West' and its civilising mission had functionally replaced the older, (openly) racist one of the 'White Man's Burden'. A key concept in this process of a second rise of the West was the notion of modernisation.

7.4 The Soviet East as the Northern Prometheus in the Second Rise of the West

After the Second World War, with old Europe's power massively diminished, a new world order emerged and a massive decolonisation process started. In particular in the 'Third World', but also in countries as Italy, still many people believed in an imminent Soviet-led world revolution. And, as pointed out above with reference to Mao, the notion of 'the East' could now signify the global socialist camp. However, this new world order was shaped by the confrontation of the two new superpowers, and under the leadership of the United States the new global West now again more strongly claimed for itself the role of the motor of human progress.

The new situation found its theoretical expression in two paradigms: modernisation theory in the West bloc and the theory of scientific-technical revolution (STR) in the East bloc. In the words of Odd Arne Westad, "Washington and Moscow needed to change the world in order to prove the universal applicability of their ideologies, and the elites of newly independent states proved fertile ground for their competition" (Westad, 2005, 4).

The modernisation theory served (1) to revise the classical concepts of Western modernity such as positivist evolutionism and Weberian rationalisation that had been developed before the First World War, in the light of the experience of fascism and Nazism, (2) to overcome the pessimistic or crisis narratives of interwar Western social philosophy and cultural critique, and (3) to develop a theoretical critique of Marxism as well as an alternative practical doctrine for the new independent nations. According to Calhoun (2007), this effort was significantly advanced by the regional studies disciplines (covering Eastern Europe, Latin America, Africa, China, and other parts of the world) that had emerged in the US during the war. However, the theoretical aspects of this intellectual endeavour (i.e. a new framework for understanding Western development) and the practical ones (i.e. an explanatory model for processes unfolding in developing countries) were not easy to reconcile (Sztompka, 1993).

The formation of classical modernisation theory took place in the 1950s and early 1960s, first in the form of Parson's structural functionalism and systems approach. Two key principles—the systemic nature of social changes and recognition of the 'unparalleled adaptive capacity' (Parsons, 1966) of modern Western societies (particularly the United States)—lead to the conclusion that the combination of social changes established in Western societies represents a universal model of social development (market economy, industrialisation, liberalism, rationalisation, and so on). Since the evolutionary approach had limited political applicability for the 'developing nations' to quickly mimic the ideal type, a stronger emphasis on the role of political and intellectual elites was needed, such as in Edward Shils' explanation of decolonisation as a dissemination of the politico-cultural universalism of citizenship (Shils, 1975). Thus, the rational efforts of the elites—guided by, or in competition with the Western model—could foster modernisation, eliminating backward traditional structures. In techno-determinist, technocratic accounts, industrialisation was seen as fostering global convergence: "As industrialism advances and becomes increasingly a world-wide phenomenon...the range of viable institutional structures and of viable systems of value and belief is necessarily reduced. All societies, whatever the path by which they entered the industrial world, will tend to approximate, even if asymptomatically, the pure industrial form" (Goldthorpe, 1971, 268). In an economic version of modernisation theory, Walt Rostow's concept of the 'stages of economic growth'– as an alternative to Marxist theory of formations, and in line with Simon Kuznets' general idea of gradual (non-revolutionary) redistribution of welfare in developed industrial capitalism—was emphasising technological development, economic growth rates, production structures, and similar factors in a linear model.

Although theories of convergence were not accepted in the East bloc, a convergence of theories paradoxically occurred in certain respects. From the 1960s on, the theory of Scientific-Technological Revolution (STR)—characterised in the West as perhaps the most important ideological extension of Marxism-Leninism since the early days of the USSR (Hoffmann, 1978)—was developed in Soviet philosophy and sociology of science and technology. At the period of an active expansion into the South during the Cold War, STR scholars faced a similar challenge as their Western opponents: to explain the opportunities for accelerated development and political agency in the 'Third World' within a universal model of progress. This was reminiscent of Russia's own leap into socialism despite its technological backwardness and underdeveloped capitalist relationships. Initially, Lenin's slogan of 'catching up with and surpassing the capitalist countries' presupposed that the best scientific-technical achievements of the West could be most effectively developed on the basis of scientific Marxism. The socialist adaptation of Taylorism and Fordism would, for example, realise the full potential of labour productivity. First concessions, then technical import, reverse engineering, copying and learning from the West were crucial in the first decade of socialist industrialisation. From 1923 to 1933, approximately 200 agreements on foreign technical assistance and supply of equipment in the heavy, steel, chemical and other industries were concluded, mainly with American and German companies, but also with France, Sweden, Italy, Switzerland and other Western countries, including a full-scale technology transfer. The Soviet side would receive full descriptions and specifications, patents and training of engineers and workers. In the early 1930s, Stalin estimated the lag behind technologically advanced countries at 50–100 years and aimed to overcome it in just ten years: "What remains for us to do is just this: to learn technology and to master science" (cit. in Buchholz & Blakeley, 1979). The pre-war decade was thus dedicated to increasing the technological autarky.

As Khrushchev recited the 'catching up and surpassing' phrase in the late 1950s, it acquired a more global character: as cited in the 1961 programme of the Communist Party of the Soviet Union: 'The ideas of Communism gain the minds and hearts of the great masses not through war with other countries but through the example of a more perfect organisation of society, through increase in the forces of production, and through creation of everything that man needs to be happy and well' (Buchholz & Blakeley, 1979). This has been characterised as a kind of 'economic ideology': "[p]roducing a faster and better modernity would allow for the final victory of Soviet socialism over Western capitalism" (Iandolo, 2012, 703f.).

Starting with a few short books and dissertations in economics in the USSR, and more extensive publications in East Germany, by the end of the 1960s the notion of the STR had gained attention in Soviet social sciences. The 1970s witnessed a boom of publications discussing the dialectics of social development on the new scientific-technological basis, including the problems of the developing countries. Despite there was no unified definition of it in Soviet literature, the general vision was expressed in a publication on humanity, science and technology in the early 1970s: "In its essence, the scientific-technological revolution expresses the laws of development of the epoch of transition from capitalism to communism, being subjected to them, and is an element of this global process in the sense that it establishes historically the material-technical preconditions for a communist society and that it can only be fully realised and utilised under socialism" (Hoffmann, 1978/1973, 618f.)

Soviet authors had to distance themselves both from Western scientism and technocracy and from conservative or leftist technophobia (Kovalev & Kovalenko, 1972). Taking a critical stance towards the theories of the 'second industrial revolution' and post-industrialism, it was important to show that STR, although starting later in the socialist bloc, would lead to the material-technological basis for the transition to communism. The famous Soviet Philosophical Encyclopedia (1960–1970) characterised STR as the major arena of the historical competition between socialism and capitalism, at the same leaving space for interaction with both the 'First' and the 'Third' Worlds. The global character of STR would require imperatively the development of international science and technology cooperation, including with countries with different political and economic systems.

The competition between the blocs for the influence on the decolonised countries led to the adaptation of Soviet internationalist discourse to the different 'Third World' contexts. Ideological integrity was possible wherever the USSR relied on the local left forces or was capable of cultivating them. However, in case of non-socialist or non-aligned states, the strategy would be to accentuate technological progress for accelerated economic growth. On the other side, Soviet analyses avoided technological determinism by accentuating the negative effects of the capitalist STR on the newly independent states, e.g. the worsening market conditions for raw materials due to competition from synthetic substitutes and reduced material intensity of production in developed nations (Shpirt, 1973). The shift from passively suffering these effects to actively harnessing the achievements of the STR, it was argued, depended on cooperation between the new nation states themselves as well as with the socialist bloc.

While the Soviet space program symbolised the futurist nature of socialism, the more 'down-to-earth' projects of electrification, railways transportation, irrigation and other rural and urban infrastructure showcased the clear and tangible positive effects of cooperation. The images of the Central Asian Soviet republics' growth with the technologisation of agriculture and urbanisation were powerful for both the internal and international audiences (Rupprecht, 2012). Thus, in the new imaginary of the world map, the 'young nations' or 'Third World' formed the new South to which the Soviet Union related itself and which was associated with the latter's 'inner East'.

The three primary methods of Soviet engagement with the expanded 'South/East' had been developed within the Soviet-Chinese cooperation: large infrastructural projects (e.g. railways), constructing the scientific-technical national curriculum, and educating cadres in the Soviet Union. From 1955 to 1960, for example, the Soviet government agreed to help build 156 facilities in the People's Republic of

China, mainly heavy industry and power plants, and some defense industry projects (Zhihua, 2000). By 1953, three quarters of the universities and colleges across the country had been adjusted to the Soviet curriculum, invited Soviet lecturers and translated textbooks from Russian to Chinese (Zhang & Jiang, 2018). Before the Sino-Soviet split, Chinese students constituted 10% of all foreign students in the universities of USSR, mostly concentrated in STEM.

The second large counterpart of the USSR in Asia, India, initiated technological cooperation in the mid-1950s with a focus on heavy, extractive and chemical industry and energy—just as the Soviet Union had one in its First Five Year Plan. Unlike China, India had both a private sector and earlier linkages with Western business, which allowed using comparative advantages. In this respect, Soviet contractors could win or lose competition in terms of level of technology, costs, control, non-restrictive transfer of knowledge (patents) and reliance on local production (D'Mello, 1988), but also win in the more subtle competition for the 'minds and hearts'—by distancing themselves from colonialism and racism: Soviet engineers "had a clear understanding of the political nature and multi-layered meaning of development projects that Western businessmen were clearly missing" (Trecker, 2024).

In the early 1960s, the Peoples' Friendship University in Moscow was founded, in the first cohort enrolling 179 Africans, 231 Asians, 182 Latin Americans, and 60 students from the Middle East. It became a long-lasting symbol of Soviet soft power in the 'South', although the majority of students were later coming to other universities all over the Soviet Union, from Kyiv to Tashkent (Katsakioris, 2017). Until the 1980s, this was the main form of regular encounters between the local population and foreigners, which certainly affected the public perception and imaginaries of the Soviet 'friendships'. The widely publicised generous aid to the South was an inspiring idea for the Soviet youth and a symbol of professionalism, especially in academia, engineering and medicine, but could also spark discontent, especially amid scarcity. At the same time, the problem of efficient formation of the pro-Soviet elites was objective. Since Soviet affirmative action and support of local political activists was opposed by many governments in the South, some of the incoming students were from well-positioned families, male and oriented towards STEM and medicine. Despite of many such technocrats eventually ending up in the political and administrative circles of their countries, it would be hard to talk about full-scale socialist indoctrination or formation of elites, especially with the students' general disinterest in political and social disciplines. This, again, mirrored the situation in the 'inner East' of Soviet Union: "Meant to accelerate the Africanization or Arabization of elites, the mission had, at least in the Soviet imagination, parallels with *korenizatsiia* (indigenization) in the Soviet republics and national territories. The prospering republics of Soviet Central Asia should serve as a model for the developing world. At international forums, Soviet delegates never failed to stress the relevance of the Soviet experience. Away from the forums, inside Komsomol and faculty offices, the mood was less optimistic. And skepticism about the prospects of creating a socialist intelligentsia out of 'bourgeois nationalists' quickly became widespread" (Katsakioris, 2017; 545).

Unlike academic exchanges which entailed some tensions, the Soviet infrastructure projects in the South were perfectly in line with the imaginaries of 'The Great Construction of Communism', especially when both exoticised and romanticised in public culture. In the 1970s, the Soviet government turned to technical interaction with non-socialist Latin America countries through large credits, alongside with equipment and personnel, for the construction of, for example, hydroelectric power plants in Argentina, Brazil, Mexico, and Colombia, irrigation projects in Peru and Venezuela, and a dam in Uruguay. In Africa, the first decade of cooperation was devoted mainly to geological exploration, and later, construction of power plants and factories in, for example, Angola, Algeria, Benin, Ethiopia, Ghana, Guinea, Mali, Morocco, and Tanzania (Kochetkova et al., 2017). Some of the projects, started during the Soviet era, have been finished only in the beginning of the 2000s.

There is no doubt that the remnants of these educational, technical, economic and military networks (Kochetkova et al., 2017; Hu et al., 2023) are still a source of the influence of Russia and, to a lesser extent, other former Soviet republics, in the Global South, despite the drastic changes in their ideologies since 1991. They also constitute a popular motive of nostalgia and pride among a large share of the post-Soviet population and probably contribute to the again relatively high prestige of STEM professions that had lost much of their social standing and resources during the neoliberal transition in the 1990s. These professional, ideological and national motives form a vague complex of collective feelings, which may be characterised as 'Prometheus defeated'. The world-historically new East, which increasingly embodied the future above all through its scientific, technological and engineering achievements, has shrunk to a nostalgic image of a progressive past. But it has left clear material traces in many places in the world and also in the revolutionised 'Far East', in China, Vietnam and elsewhere, and it has (co-)initiated developments that still today call into question the globally necessary character of the Western path of modernisation.

7.5 Conclusions Responding to Mitcham's "Re-Imagining Chinese Technological Imaginaries"

Soviet progressivism presents a difficult case in terms of imaginaries of the self and of the 'others', which in different contexts included both enchantment with and resentment to the West, as well as vague inner and outer Orientalism. The cult of progress that was rooted neither in Protestantism nor in Enlightenment, but took its radical forms in Russian Marxism, and in the quasi-religious Cosmism, included in itself the motives of self-construction and self-transcendence. In comparison to the early merchant maritime cultures, or European overseas colonialism, the Russian Empire's and then the Soviet expansion within the continent created the peculiar relationship with space and with the bordering nations—a kind of an ever-changing imaginary geography, with the West-East-North-South distinctions relativist and situative. The 'inner East' of Soviet Union—its vast Asian territory, largely uninhabited, with a scattered population—appeared as a blank slate, a space where the transfer of ideas, technologies, resources and people was perceived as a perpetual possibility. This 'great construction project', entailing immense effort, losses, and displacement, resonated with universalist thinking—the idea of a 'world revolution' was reinterpreted in the form of a scientific-technical revolution. STR discourse embodied the hope that a non-Western conception of technoscientific and engineering progress would prevail globally and even in the end lead to an overcoming of capitalism. While this hope in the end did not materialise, it reversed in a powerful imaginary the centuries-old Western narrative described by Adas and others. In some respects, the global activities of the Soviet engineers can be seen as 'translation' of socialism into technology and of technology into socialism. In a paradox way, the general result of this process was a politicisation of technology along with 'neutralisation' of socialism. As the agents of change, however, the Soviet engineers fulfilled the subsidiary function of "peoples' diplomacy", personifying the Soviet mission in the world.

As we have tried to show, Russian revolutionary activities in the first fifth of the twentieth century strengthened hopes for a new East in the universalist socialist working-class movement worldwide. While this East still viewed Western societies as a model in some respects, particularly in science, technology, and engineering, it contradicted the Western narrative of progress and modernisation. And it included the claim that this new East was a kind of better. More modern West and fulfilled the universalistic, liberating, and Promethean hopes that the West, which now seemed old and reactionary, had betrayed.

At the same time, it became clear in these years that culturalist metaphors can be very dangerous. As Luxemburg pointed put in her influential text 'The Russian Revolution' in 1918, the anti-tsarist and anti-Russian lying phrases of right-wing and mainstream Social Democrats glorifying the German bayonets had a very real meaning and fatal consequences in the history of the class struggles of these years. The metaphor of 'the East' was used to justify an imperialist war against old Russia, ignoring that there existed a new Russia that should have been the focus of the emancipatory forces in the West. We have tried to show that, at least from the 1870s on, this was the persuasion and a strong hope of Marx. The reactionary use of the East-West dichotomy by German and other social democrats played a key role in the catastrophe that the First World War represented for world history. As a more general lesson, we can thus say that pseudo-geographical metaphors should always be used with great caution. The arbitrary linking of progressive and anti-progressive characteristics in such metaphors and the lack of distinction between cultural, socio-economic, political and technical-scientific aspects of progress clouded the view of the world and the emancipatory forces in the West became "dis-orient-ed": metaphorically speaking, they did not see the light from the East, the Russian Revolution.

In his lecture "Re-Imagining Chinese Technological Imaginaries" (this volume), Carl Mitcham focused his analysis—unlike ours—on China. Nevertheless, in our conclusions, we would also like to explore his remarks against the background of our own research and reflections.

When he calls the British 'imperialist barbarians' in reference to their actions in China in the mid-nineteenth century, we believe this is not simply a polemical reversal of Western stereotypes of a supposedly barbaric East. It reflects real history, including the views of many Chinese at the time. As we have tried to show, Marx also found British and Western capitalist barbarism the worst and, in many ways, most shocking. And we also agree with Mitcham when he castigates Western ignorance of Chinese history and philosophy, as well as Western Orientalism.

As argued elsewhere (Coenen, forthcoming), there is a shameful imbalance between knowledge about China in the West and Chinese knowledge about the West. From our perspective, the real history of comprehensive, i.e. technical-scientific and social progress must be reclaimed against Western ideology, and the critique of anti-'Eastern' and anti-'Southern' Eurocentrism must become a key element of materialist superstructure analysis, with aspects of the history of religion and philosophy also being considered central. Since Gramsci, the deficiencies and underdeveloped character of superstructure analysis in Marxism has often been criticised. In our work, we have tried to contribute to such an analysis with respect to metaphorical East-West dichotomies, focusing on Russia and the Soviet Union and relating metaphorical discourse to the real history of science, technology and engineering. Doing the same for China is an important task for our fields of research.

Mitcham's response to Joseph Needham's question why China fell behind the West in terms of technoscientific development—his 'thought experiment' not to present modern China as deficient with regard to the West but to take the country as a norm from which to reflect back on the West—is an inspiring attempt to make sense of the world we live in now. In our contribution, we focused on showing that already analyses by (in particular the late) Marx himself do not support the notion that something went wrong with China, or with the East more generally. Quite to the contrary, as we can also see, for example, from the analyses by Luxemburg, Western developments and in particular imperialism were what went utterly wrong in modern history, where the progress of civilisation in terms of science, technology and engineering potentiated barbarism and gave it a new quality. It appears to us, however, that even some of the best 'classical' Marxian and Marxist analyses still suffered from questionable historical determinism and unclear periodisations as well as from dubious uses of geographical metaphors and culturalist assumptions. It may be more appropriate, for example, to more closely look at the commonalities between pre-capitalist societies, as Samir Amin (2009; cf. Coenen, forthcoming) does with his concept of 'tributary cultures', with which he avoids stretching non-Western societies and cultures into the Procrustean bed of European history, with its fixation on the Roman Empire and its feudalist 'Middle Ages'. Amin's proposal to conceive of the times from Alexander the Great to the emergence of capitalism as the age of tributary culture and to explain the rise of 'the West'—and later Japan—by their originally peripheral, barbaric or semi-barbaric position in relation to the centres of tributary culture appears to us as very useful also with respect to the analysis of the modern history of Chinese-Western interactions. And we agree with Kwame Anthony Appiah (2016) that the notion of Western civilisation or culture exclusively representing human progress in terms of liberty, tolerance and rational inquiry is an empirically unfounded ideologeme. In general, as Appiah argues, cultural identities can have limiting and repressive as well as emancipatory features but "our lives must make sense, too, at the largest of all scales. We live in an era in which our actions, in the realm of ideology as in the realm of technology, increasingly have global effects. When it comes to the compass of our concern and compassion, humanity as a whole is not too broad a horizon" (Appiah, 2016, n.p.).

Mitcham's critique of what he sees as fundamental Western deficiencies caused by historical, and in particular religious and mythical roots of Western culture—its devaluation of rural and everyday life, of the embodied pleasures and perennial sorrows of in particular female domesticity, and peacefulness, its highly conflict-oriented understanding of the world, the focus on war, commerce, exploration and conquest, and the discontinuity of and internal tensions within the Western history if ideas—, is consistent with Marxist analyses of modern history that avoid glorifying the destructiveness of Western capitalism and with Amin's analyses of originally peripheral Western European culture. However, the question can be raised whether China's renaissance since the 1940s was not based at least as much on the Chinese appropriation of perhaps the most dynamic and conflict-oriented current of Western thought, namely Marxism, as on the strength of Chinese premodern cultural traditions And if one shares the view that a significant part of modern reformist and (liberal-)democratic Enlightenment thought was strongly influenced by the reception of Confucian ideas (Lee & Rüdiger, 2020), one could even argue that intellectual traditions from the 'Far East' have indirectly influenced socialist democratic thought in the West (including Germany and perhaps also Russia).

Last but not least, we would like to point out that Mitcham's thought experiment, which focuses on the role that religious and mythical roots of Western self-understanding have played and continue to play in the West's aberrations, is indeed a thought-provoking and highly inspiring proposal for science, technology and society studies. As argued elsewhere (e.g. Coenen, forthcoming), there is an irritating tendency in Western thought to load science, technology and engineering with metaphysical fears and, to a lesser extent, hopes. An interesting research question for the future could be the analysis of East-West dichotomies in the context of the metaphors of light and darkness, which have strongly influenced Western thought from antiquity through the Age of Enlightenment to the present day. Above all, however, Mitcham's observation that Chinese imaginaries are at home in the world, while Western ones are characterised by a longing for a supernatural home, restlessness, an obsession with overcoming boundaries, and a profoundly unpragmatic view of science, technology, and engineering, deserves further attention.

References

Adas, M. (1989). *Machines as the measure of men: Science, technology, and ideologies of Western dominance*. Cornell University Press.

Amin, S. (2009/1988). *Eurocentrism* (2nd ed.). Monthly Review Press.

Appiah, K. A. (2016, November 9). There is no such thing as Western civilisation. *Guardian*. https://www.theguardian.com/world/2016/nov/09/western-civilisation-appiah-reith-lecture

Benjamin, W. (1928/1978). To the planetarium. In *Reflections: Essays, aphorisms, autobiographical writings* (pp. 92–94). Harcourt Brace Jovanovich.

Bonnett, A. (2004). *The idea of the West. Culture, politics and history*. Palgrave Macmillan.

Buchholz, A., & Blakeley, T. J. (1979). The role of the scientific-technological revolution in Marxism-Leninism. *Studies in Soviet Thought, 20*(2), 145–164.

Calhoun, C. (Ed.). (2007). *Sociology in America: A history*. University of Chicago Press.

Christoff, P. K. (1972). *An introduction to nineteenth-century Russian Slavophilism, Vol. 2: I. V. Kireevskij*. Mouton.

Coenen, C. (forthcoming). A critique of the West's understanding of technology and itself in the age of AI. In G. Wang, C. Mitcham, & A. Nordmann (Eds.), *AI-ethics: A cross-cultural dialogue*. Springer.

Coenen, C., & Kazakova, A. (2021). Utopian grammars of human-machine interaction. *Technology and Language/Технологии в инфосфере, 2*(1), 67–80.

D'Mello, B. (1988). Soviet collaboration in Indian steel industry, 1954–84. *Economic and Political Weekly, 23*(10), 473–486.

Deutscher Reichstag. (1914). Erklärung der Sozialdemokratischen Partei zum Kriegsausbruch, abgegeben vom Fraktionsvorsitzenden Haase im Reichstag (4. August 1914). *Verhandlungen des Reichstags*, XIII. LP., II. Sess. 1914, Vol. 306, 8f.

GoGwilt, C. (1995). *The invention of the West: Joseph Conrad and the double-mapping of Europe and Empire*. Stanford University Press.

Goldthorpe, J. H. (1971). Theories of industrial society: Reflections on the recrudescence of historicism and the future of futurology. *European Journal of Sociology, 12*, 263–288.

Gramsci, A. (1977/1975). *Quaderni del carcere II. Quaderni 6-11* (V. Gerratana, Ed., 2nd ed). Einaudi.

Hoffmann, E. P. (1978). Soviet views of "The scientific-technological revolution". *World Politics, 30*(04), 615–644.

Hu, Z.-A., Li, J., & Nie, Z. (2023). Long live friendship? The long-term impact of Soviet aid on Sino-Russian trade. *Journal of Development Economics, 164*, 103117.

Iandolo, A. (2012). The rise and fall of the "Soviet model of development" in West Africa, 1957–64. *Cold War History, 12*(4), 703–704.

Ienna, G., & Rispoli, G. (2021). The 1931 London Congress. *HoST—Journal of History of Science and Technology, 15*(1), 107–130.

Katsakioris, C. (2017). Creating a socialist intelligentsia. *Cahiers d'études africaines, 226*, 259–288.

Kautsky, K. (1902, March 10). The slavs and the revolution [Славяне и революция]. *Iskra/Искра* 18

Kautsky, K. (1907). *Patriotismus und Sozialdemokratie*. Leipziger Buchdruckerei AG.

Kochetkova, E., Damtar, D., Boliachevets, L., Slyusarchuk, P., &Lajus, J. (2017). *Soviet technological projects and technological aid in Africa and Cuba, 1960s–1980s*. HSE Working Papers No. WP BRP 143/HUM/2017, National Research University Higher School of Economics.

Kovalev, A. M., & Kovalenko, V. I. (1972). On the question of the interrelations between scientific--technological and social revolution. *Soviet Studies in Philosophy, 10*(4), 383–394.

Lee, E.-J., & Rüdiger, A. (2020). China und die moderne Gleichursprünglichkeit von Demokratie und Meritokratie. Zur interkulturellen Ideengeschichte des Republikanismus in der Epoche der Aufklärung. *Allgemeine Zeitschrift für Philosophie* 45(1) (2020), pp. 27–50.

Leung, J. Y. (2013). A critique of Marx's view of the Taiping rebellion and its origins. *Armstrong Undergraduate Journal of History, 3*(2), 4.

Li, B. (2019). The reason why Mao Zedong put forward the theory of "The East wind prevails over the West wind" and his relevant strategic thinking (毛泽东提出"东风压倒西风"的缘由及战略思考). *Journal of Xiangtan University (Philosophy and Social Sciences), 42*, 3. http://www.hprc.org.cn/gsyj/zzs/mzdsxyj/201912/P020191231327679775724.pdf

Liebknecht, W. (1878). *Zur orientalischen Frage oder Soll Europa kosakisch werden?* (2nd ed.). R. E. Höhme.
Lindner, K. (2010). Marx's eurocentrism. Postcolonial studies and Marx scholarship. *Radical Philosophy, 161*, 27–41.
Luxemburg, R. (2000). Die Krise der Sozialdemokratie. In *Gesammelte Werke* [RLW] (Vol. 4., 6th ed., revised, pp. 49–164). Karl Dietz.
Luxemburg, R. (2005). Die Russische Revolution 1905. *UTOPIE kreativ, 171*, 55–66.
Marx, K. (2008). *Dispatches for the New York tribune: Selected journalism of Karl Marx* (J. Ledbetter, Ed.). Penguin.
Masaryk, T. G. (1919). *The spirit of Russia* (Vol. 1). George Allen & Unwin and Macmillan.
Parsons, T. (1966). *Societies: Evolutionary and comparative perspectives*. Prentice-Hall.
Philosophical Encyclopedia [Философская Энциклопедия]. (1960–1970). *Scientific-Technical Revolution (STR)*. Soviet Academy of Sciences.
Ponce, A. (2007/1935). Visita al hombre future. In S. Sattía (Ed.), *Hacia la revolución* (pp. 125–141). Fondo de Cultura Ecónomica.
Rupprecht, T. (2012). Soviet internationalism after Stalin: Interaction and exchange in the 'Third World' during the Cold War. *Journal of Modern European History, 10*(2), 195–217.
Schulte Nordholt, J. W. (1995). *The myth of the West. America as the last empire*. Wm. B. Eerdmans.
Schöler, U. (2021). *"Despotischer Sozialismus" oder "Staatssklaverei"? Die theoretische Verarbeitung der sowjetrussischen Entwicklung in der Sozialdemokratie Deutschlands und Österreichs (1917–1929)*. Dietz.
Sdvižkov, D. (2015). Russian and German Ideas of the West in the Long Nineteenth Century. Entanglements of Spatial Identities. In R. Bavaj, & M. Steber (Eds.) Germany and 'the West'. The history of a modern concept (pp. 97–110). Berghahn.
Shanin, T. (Ed.). (1983). *Late Marx and the Russian road: Marx and the peripheries of capitalism*. Monthly Review Press.
Shils, E. (1975). *Center and periphery: Essays in macrosociology*. University of Chicago Press.
Shpirt, A. Y. (1973). The scientific-technical revolution and the development problems of the "Third World" countries [Nauchno-tekhnicheskaya revolyutsiya i problemy razvitiya stran "tret'ego mira"]. *Voprosy Istorii, 7*, 57–72.
Suslov, M. (2012). The lost chance of conservative modernization: S. F. Sharapov in the economic debates of the late nineteenth to the early twentieth century. *Acta Slavica Iaponica, 31*, 31–54.
Sztompka, P. (1993). *Sociology of social change*. Blackwell Publishers.
Trecker, M. (2024). Forging the Indian steel industry: How Soviet designs won the day. *Cold War History, 24*(4), 517–536.
Varouxakis, G. (2025). *The West: The History of an Idea*. Princeton University Press
Westad, O. A. (2005). *The global Cold War: Third World interventions and the making of our times*. Cambridge University Press.
Zhang, B., & Jiang, L. (2018). The influence of the USSR on Chinese engineering education: Taking Beijing College of Aeronautics as an example. In *Proceedings of the 2018 International Conference on Engineering Technologies and Computer Science (EnT)* (pp. 14–20).
Zhihua, S. (2000). Sino-Soviet relations and the origins of the Korean War: Stalin's strategic goals in the far East. *Journal of Cold War Studies, 2*(2), 44–68.

Christopher Coenen is a political scientist working in the field of technology assessment (TA) at the Institute for Technology Assessment and Systems Analysis (ITAS) within the Karlsruhe Institute of Technology (KIT) where he leads the research group 'Life, Innovation, Health, and Technology'. He is editor-in-chief of the journal "Ethics and Society" (SpringerNature).

Aleksandra Kazakova is a researcher in Engineering Studies. She was a senior lecturer in Contemporary Social Theory and Sociology of Technology at Lomonosov Moscow State University, Bauman Moscow Technical University and Gubkin Russian State University of Oil and Gas (2012–2022), a visiting researcher at the Institute for Technology Assessment and Systems Analysis within the Karlsruhe Institute of Technology (2018–2019) and at the STS Center of the European University at St Petersburg (2021–2022), and a PhD candidate in Philosophy of Science and Technology at the University of Chinese Academy of Sciences in Beijing (2022–2025).

Chapter 8
Salvation by Technology: A Note

Sajay Samuel

Abstract In the 1990s, Ivan Illich hypothesized a constitutive relationship between Latin Christian theology and the concept of technology as we understand it today. Recent scholarship by Giorgio Agamben and Nicolas Heron gives credence to the proposition that technology is a historical phenomenon. In this research note, I outline Illich's hypothèses, summarize the relevant literature substantiating his claims, and list some open questions about the relation between theology and technology. This note proposes that the belief in "salvation by technology" animates Western techno-scientific civilization.

Keywords Ivan Illich · Theology and technology · Instrumental cause

8.1 A Question Concerning the Instrument

The phrase "Guns don't kill people, people kill people" has become a cliché after long use by the National Rifle Association of the United States. It is not to debate the merits of gun ownership that I resort to platitude. Instead, it is to convey the dominant idea of technology in Western culture. The popular understanding of technology is of a designed artifact that conveys the intention of its user. A gun is designed to kill. Whether it is used for sport or murder depends on the intention of the user. Two interrelated features are integral to this conception of technology: that it has a built-in purpose distinct from that of its user and that it conveys the intention of its user without distortion. It is in respect of these two features—distality and neutrality—that cars, schools, and medical protocols are considered as of the same kind. Each is a technology because it is an artifact or convention designed to accomplish a particular task but whose specific use is determined by the user. This

S. Samuel (✉)
Pennsylvania State University, University Park, PA, USA

L. Checketts (ed.), *What Went Wrong with the West: Carl Mitcham's Critique of Western Philosophy of Technology*, Philosophy of Engineering and Technology 52, https://doi.org/10.1007/978-3-032-22214-5_8

instrumental understanding of objects and conventions is widespread, though recently disputed by some scholars of science and technology studies (Sismondo, 2010).

In his posthumously published discussion with David Cayley, Ivan Illich credits his conversation with Carl Mitcham for the two novel hypotheses he presents (Cayley, 2005, pp. 72–75). Illich sketched these hypotheses in a two-step argument. First, he asserts that the idea of technology has a history. By insisting on its historicity, Illich upends the unexamined assumption that technology is of a natural kind. For instance, that is what primatologists presume in positing that chimpanzees use branches as tools when foraging for ants, or what anthropologists suppose in discovering tool use by aboriginal Australians, or what historians of technology believe when discussing the Aztec *tecpatl* as an instrument for human sacrifice. Imputing technology to non-Western peoples or animals is mistaken because the concept, as we understand it, emerged in the West. In fact, neither the ancient Greeks nor the Romans had our idea of technology, though they built aqueducts, wielded swords, and used maps. Accordingly, Illich noted that, for Aristotle, the word *organon* (Ὀργανον), which is usually translated as "instrument," referred to both "my hand without the pencil and my hand armed with the pencil" (Illich, in Cayley, 2005, p. 73).

In Illich's view, this linguistic ambiguity between the hand and the tool-holding-hand reveals that Aristotle did not conceive the distality implied by our concept of technology. To remain with his anachronistic example, the pencil was not conceived as a thing capable of writing and available to transmit the intention of the user—whether to write a poem or underline a passage. The mental frame in which the pencil is not perceived as an instrument is foreign to us. We betray our provinciality in denying the historicity of technology.

The absence of the concept of the instrument, for example in Aristotle, led Illich to explore when and why the idea of technology emerged in the West. His second hypothesis was that medieval Christian theology constitutively shaped the concept of technology. In particular, he suggested that "a world conceived in the spirit of contingency [lay] at the origin of the modern conception of tools" (Illich, in Cayley, 2005, p. 75). Illich accepted the claim by Hans Blumenberg that the notion of contingency was a peculiarly Christian idea. From Augustine to Aquinas, the idea of contingency implied "a society in which people had strongly experienced the world as lying in the hands of God" (Illich, in Cayley, 2005, p. 69). Contrary to the Greek argument that the world was eternal, Christians, particularly after Augustine, believed that God made the world from nothing (*ex nihilo*). Accordingly, the existence of the created world rested wholly on the will of the creator. Hence, contingency "was a conception that saw…a universe that would disappear if [God's] hands disappeared, and which is necessary only insofar as it depends on his will" (Illich, in Cayley, 2005, p. 74). Contingency in this sense does not imply chance or luck. Rather than random occurrences, contingency in the medieval period expressed the feeling that all things great and small depended on God's loving grace.

A universe that depends entirely on God prompts the question of how it is governed. And it is in Aquinas' answer to this question that Illich finds the origins of our idea of technology. Aquinas argued that angels had instruments—the heavenly bodies—through which they mediated God's will to govern the world. Equally, says Illich, theologians argued for sacraments as instruments of God (*instrumenta dei*). In this way, the uniquely Christian idea of the world lying in the hand of God prompted the concept of the instrument as the mode by which God's work on earth was done. However, it was only when "the world fell from the hands of God into the hands of man…that…the way opened for a fully technological society" (Illich, in Cayley, 2005, p. 76).

Scholars—historians, sociologists, or philosophers—concerned with the study of science and technology have almost completely ignored these twin hypotheses. Instead, it is in the work of two genealogists of biopolitics, Giorgio Agamben and Nicholas Heron, that one finds an in-depth study of the historicity of technology, which affirms its constitutive relationship with Latin Christian theology. While Agamben explicitly acknowledges and develops Illich's original insight into the theological roots of technology, Heron's study offers extensive context to substantiate Illich's hypotheses.

8.2 Towards a Genealogy of the Instrument

Giorgio Agamben is the preeminent genealogist of biopolitics. His multi-volume work on the subject ended with a ninth book in which he exposes the historicity of the instrument as such (Agamben, 2016). In recognizing the path forged by Illich before him, Agamben confirms that "the discovery of the instrumental cause is the first attempt to give a conceptual figure to technology" (Agamben, 2016, pp. 72–73).

However, the concept of technology did not emerge fully formed in the thirteenth century, as did Athena from the head of Zeus. The historian Lynn White has noted the twelfth century was a period of remarkable innovation and diffusion of technical devices (White, 1962). Moreover, as Illich points out, at least two books published around 1125 hint at the idea of technology (Illich in Cayley, 2005, p. 73). Theophilus Presbyter, a Benedictine monk and practicing metalworker, published *De Diversis Artibus*, a compilation of devices and techniques for painting, illuminating texts, producing stained glass, building church organs, and goldsmithing. What is notable about this handbook for craftsmen is that the depictions of diverse objects and techniques are all presented together as if of one kind. Though the idea of the instrument is not yet elaborated, the compilation of diverse artifacts and technical processes suggests the recognition of their similarity. About the same time, Hugh of St. Victor published his *Didascalion,* which is acknowledged as the first time the practical or mechanical arts are brought within the fold of science or theoretical knowledge (Illich, 1981, p. 90). As such, he can be understood as presuming the categorical existence of devices before these are conceptualized as instruments.

The concept of the instrument is beholden to the idea of an instrumental cause, which was unknown to Aristotle. In the Aristotelian scheme, the explanation for a phenomenon required identifying its four causes. The pot is explained by the concatenation of its material cause (clay), its formal cause (shape), its efficient cause (potter), and its final cause (purpose of storing food). In such a scheme, there is no place for the potter's wheel, understood as an instrument that transmits the intention of the potter to the clay. Rather, the potter's hand subsumes the potter's wheel. It is obvious in considering Aristotle's remarks on the relationship between the artisan and the tool that he does not conceptualize the instrument as such. In the *Eudemian Ethics*, Aristotle argues that:

> the relations between soul and body, artisan and tool, and master and slave are similar, between the two terms of each of these pairs there is no partnership (*koinonia*); for they are not two, but the former is one and the latter is part of that one, not one itself; nor is the good divisible between them, but that both belong to the one for whose sake they exist. For the body is the soul's tool (*organon*) born with it, a slave is as it were a member or tool (*organon*) of his master, a tool (*organon*) is a sort of inanimate slave (1241b).

The presumption that the relation between artisan and tool is not external and accidental but internal and essential startles us. Why "being a part of" something prevents a partnership with it is almost incomprehensible today. Agamben focuses on the master-slave relationship to clarify the meaning in Aristotle of "being a part of" and thereby explains the emergence of the instrument as such (Agamben, 2016, pp. 12–14). Aristotle says there can be no partnership between the master and slave because the latter is "part of" the former. We automatically think that the slave is "a part of" the master because he is owned by the latter. However, for Aristotle, "being part of" does not imply the juridical relationship of property between master and slave. Instead, "being part of" entails *participating in*, just as the body participates in the soul. For Aristotle, the slave is an "animate instrument" (*organon*) of the master, but this *organon* should evoke, for us, the sense of integral body organs rather than separable external instruments. To clarify this "organic" sense of the Aristotelian *organon*, Agamben says, "the use of a slave's body is similar to that of a bed or clothing…" (Agamben, 2016, p. 12, emphasis removed). We have been so instrumentally shaped that we automatically think of a bed as a tool for some ulterior purpose—to rest, sleep, and so on, whereas for Aristotle a bed is simply used, in the course of daily life.

Instruments are absent in Aristotle. But that is not true of Thomas Aquinas, who conceptualizes the instrument in the context of explaining how the sacraments work. Salvation is conferred by the grace of God, which is effected through the sacraments (ST, Part III, Q 62). In elaborating how the sacraments convey God's grace, Thomas distinguishes between the principal cause and the instrumental cause and thereby defines the instrument as such.

God is the efficient cause of his grace. To explain how God's grace reaches the world, Thomas must connect this efficient cause to its effect. To do so, Thomas decisively explodes the Aristotelian schema by subdividing the efficient cause into two: the *agens* or *causa principalis* and *agens* or *causa instrumentalis* (Agamben, 2016, p. 70). Whereas the principal cause refers to the actions of the user of the instrument, the instrumental cause refers to the actions of the instrument. Accordingly, Thomas defines the instrument as having

> a twofold action: one is instrumental, in respect of which it works not by its own power but by the power of the principal agent; the other is its proper action which belongs to it in respect of its proper form: thus, it belongs to an ax to cut asunder by reason of its sharpness, but to make a couch insofar as it is the instrument of an art. But it does not accomplish the instrumental action save by exercising its proper action: for it is by cutting that it makes a couch (ST, Part III, Q 62, art 1, reply 2).

The instrument is distinct from and subordinate to the principal agent. Thomas conceptualizes the instrument by conferring an active power to tools—the axe to cut—whose specific use is determined by the artisan. Technology is thereby given the conceptual form that is familiar to us: of a thing (X) whose proper action on something (Y) occurs in service of or at the behest of another (Z).

This is the precise sense that we perceive what Illich called the "distality" between the instrument and the user. Crucially however, this notion of instrumental cause also means that the ax is at once both cause and effect (Agamben, 2013, p. 53). On the one hand, it is the ax that causes the cut on the tree (through its proper action). On the other hand, it is the woodsman who causes the ax to move in the manner needed to cut the tree (through its instrumental action). The cut is caused by the ax whose proper action is caused by the woodsman. Accordingly, the instrument is that thing which acts only because it is used and is used only insofar as it acts. Otherwise said, the instrument only acts in its proper action when activated by the user. It is in this sense that technology neutrally conveys the intention or action of the user. The dominant understanding of technology today as distinct from the user and yet only conveying the intention of the user—distality and neutrality—thus finds its genealogical roots in theology.

8.3 Salvation by Technology

To elaborate the theological context for the emergence of the instrument is not possible in a short compass. As Agamben notes, the theological paradigm of technology requires understanding both "the Trinitarian economy and the doctrine of sacraments [that] are its eminent *loci*" (Agamben, 2016, p. 72).[1] Even a detailed description of only the sacraments as exemplary instruments would require more than can be offered here. I only highlight as much of the theology of sacraments as needed to appreciate the constitutive relationship between theology and technology.

Recall that Thomas brings the instrument into view by recognizing its causal power. Thus, the instrument is a device or procedure whose intrinsic efficacy is applied to one thing at the behest of something else. Heron (2018) explains what he names "liturgical power" as the effort to structure action, such that actions by X on Y are done in the name or service of Z. Hence, for instance, liturgical power is exemplified by the activities conducted by the clergy on the laity in the name of

[1] Agamben (2013, 2016) and Heron (2018) are foundational to a fuller understanding of the relationship between theology and technology.

God. Thus, the instrument and liturgical power are formally comparable as modes of vicarious action.

The "minister" is the linchpin to understand the coincidence of the instrument and liturgical power. When Thomas speaks of the "minister" as an "animate instrument" he is consciously invoking the slave in Aristotle's account of the *organon* (Agamben, 2016, p. 75). However, whereas the instrumentality of the slave lies in being used as part of the master, the instrumentality of the minister lies in his willing subjugation to God. The essential theological context for the emergence of technology lies in how the minister *intends* to work as an instrument of God (Heron, 2015).

Thomas elaborates the idea of the instrument in the context of answering how the sacraments effect God's will in the world. Ministers and consecrated oil are examples of instruments needed to mediate divine government. Thomas suggests several reasons why an omnipotent God needs to be mediated. A retinue of helpers frees God from engaging in mundane tasks; allows his power to be amplified through intermediaries; serves to display his grandeur; and, as instrumental causes, perfect the execution of divine providence (Heron, 2018, pp. 83–4). Thomas justifies the clergy being separated from the laity because they belong to a higher rank of perfection. The clerics bring the laity towards the likeness of God, insofar as he is good. In so doing, ministers imitate God, not necessarily by being good themselves, but by being the cause of good in others. The clergy are therefore instruments of God who serve the flock—they are God's administrators insofar as they minister to the laity (Heron, 2018, p. 84).

To explain how God's will is enacted in the world, Thomas instrumentalizes divinity itself.[2] He distinguishes between a conjoined instrument (*instrumentum coniunctum*) such as the potter's hand and separated instruments (*instrumenta separata*) such as the potter's wheel. He affirms the Trinitarian argument that Christ, the savior of the world, is a single person in whom two natures are united. However, he argues that the humanity of Christ is a conjoint instrument of his divinity (*humanitas instrumentum divinitatis*), just as the body is to the soul. By construing Christ's humanity as an instrument of his divinity, Thomas introduces the instrumental cause as a sub-species of the principal cause. But, since they are conjoined, Christ's humanity cannot distort or dilute the efficacy of Christ's divinity. This concept of an undistorting mediator furnishes Thomas with a model for the sacraments as instruments that effect grace (Heron, 2018).

Consider as an example the sacrament of baptism in the Catholic Church. The pouring of water in combination with the utterance of a specific set of words and actions conveys to the baptized six supernatural graces, including that of being cleansed of sin. The material qualities of water—smell, color, purity—are irrelevant to its capacity to receive and transmit divine grace. Polluted waters can wash away sins with the same efficacy as clean water because "the power to bestow salvation flow(s) from the divinity of Christ through his humanity into the sacraments themselves" (Aquinas, in Heron, 2018, p. 106). It is precisely its status as an instrument

[2] This section relies heavily on Samuel (2023), pp. 265–266.

that allows water to cleanse regardless of its material qualities. If water can cleanse because it is an instrument of God, so should the minister, but with one difference. Thomas had to contend with the difference between inanimate and animate instruments.

Whereas water is an inanimate instrument, Thomas defines the minister as a living instrument (*instrumentum animatum*) or, more precisely, as an instrument with intelligence (*minister est sicut instrumentum intelligens*). Unlike water, which must be moved in order to move, men can move autonomously; they are "automobile." Unlike inanimate water, which cannot not receive and transmit the saving grace of God, the animate instrument can disobey, rebel, or refuse. Thomas confronts this possibility by yoking the minister to the will of God. He stipulates that the minister follows God's will by acting in a ministerial way (*per modum ministerii*).

A minister acts in a ministerial way insofar as "he intends to do what Christ and the Church do" (Aquinas in Heron, 2018, p. 90). To do as Christ does, the minister must wholly subject himself to that principal agent (*se subjiciat principali agenti*). To be wholly subject to the principal agent requires the minister to be an undistorting conduit of Christ's activity. To serve as an unimpeded passage for the principal cause, the minister must function as an inanimate instrument. It is to the diminution of human potentiality into an instrument that Thomas' notion of *intentio* refers when he says the minister must "intend to do what Christ and the Church do." Crucially therefore, *intentio* "names nothing subjective, and admits of no qualification whatsoever. Neither can it be considered right or wrong" (Heron, 2015, p. 139). For Thomas, someone merely going through the motions expresses *intentio*. Yet, for us, such habitual behavior and reflex actions would not be considered intentional in the ordinary meaning of that word. It is for this reason that Ivan Illich described instrumental cause as a "cause without an intention" (Illich, in Cayley, 2005, pp. 71–79, but Cf. Heron, 2018, ff.7, p. 172).

Intentio only implicates the work done and not the worker. Just as both clean and dirty water can wash away sins when used in the baptismal rite, so also a wicked priest or layman can successfully baptize by merely uttering the right words. The sacraments work because their efficacy depends only on the work done (*ex opere operato*) and not on the effort of the worker (*opus operantis*). That is, sacraments "effect what they signify" (Aquinas in Agamben, 2013, pp. 52–53). Just as inanimate water cannot avoid being an effective agent of the principal cause, so also a human who works in a ministerial mode cannot but effectively convey the grace of God. Thus, the sacraments and priests serve as the paradigm for technology, understood as the domain of separable and neutral things, conventions, or persons that work on someone or thing at the behest of someone else.

Agamben underscores another point that is crucial to any discussion of the relationship between theology and technology. By the seventeenth century, the Jesuit philosopher and theologian, Francisco Suarez argued that, for divine instruments, "the action connatural to the instrument precedent to the action and to the effect of the principal agent is not necessary" (Agamben, 2016, p. 76). For Suarez, instrumental causality is irrelevant to divine instruments because their "obediential potential" is sufficient to secure the effects of grace given the unlimitedness of divine virtue. Suarez thereby also renders irrelevant the theological arguments developed by Thomas in support of the sacraments as instruments of God.

The eclipse of instrumental cause in Suarez finds resonance in early modern accounts of causality, by which time the scheme of Aristotelian four-fold causality has given way to an understanding of causality in which the efficient cause is only one. In this way, the efficient cause reabsorbs what once was split from it as its subordinate—the instrumental cause. As in the phrase "guns don't kill people, people kill people," the causative power of the instrument disappears only by absolutizing their efficient cause, which is man.[3] Conversely, the overshadowing of instrumental cause since the early modern period also lends novelty to its rediscovery, by not only Illich but also by the dominant currents of science and technology studies today, exemplified by Bruno Latour.

To summarize, Thomas conceptualizes the instrument in the context of developing a theology for the sacraments that convey God's saving grace. The overarching purpose of sacraments is to be the instrumental cause of good in others. Sacraments are signs that effect the salvation of man by conveying the power of the principal agent. Whether animate or inanimate, sacramental instruments in the ecclesiastical hierarchy offer the conceptual form of technology as such. Lewis Mumford should not have dismissed Buckminster Fuller as "that interminable tape-recorder of 'salvation by technology'" (Mumford, 1975, p. 44). Perhaps Fuller had correctly intuited the fundamental animus of techno-science in the West.

8.4 Some Questions

In a recent essay, Levi Checketts expressed hope that the "theological studies of technology" might flourish as an academic discipline (2022, p. 140). The foregoing theological history of technology poses at least two clusters of questions needing historically nuanced answers to this emerging field.

First, it suggests how there can be *both* Christian religious support for *and* objection to technology. The instrumentalization of salvation through the sacraments suggests the nexus between Christianity and technology. It makes comprehensible the claim of Fudpucker: "Technology not only comes forth from Christianity, it takes us into Christianity in a new and fuller sense" (2022, p. 40). Indeed, David Noble (1999) has carefully documented precisely such a "religion of technology" in his trenchant essay on how Christian millenarianism encouraged and continues to fuel advancements in techno-science. In contrast, in his masterful review of the

[3] Consider the following definition of technology offered by Checketts: "I define technology as something (whether a physical tool, a social system or even new organisms) created by human beings to channel power, typically to accomplish some end (though the success of attaining a particular end varies)" (2022, v.1, p. 94). Whether "power" refers to the principal cause (human beings) or the instrumental cause (of the physical tool, etc) is not clear. Even if the word refers to both, the resulting conflation implies the return of the efficient cause alone.

considerable research on the relationship between theology and technology, Borgmann concludes that much of that work shows "the need to see Christianity and technology as adversaries—not simply opponents, but as forces that confront one another at the deepest level" (Borgmann, 2022, p. 215). In the light of this sacramental history of technology, it is the Christian objection to it that demands explanation.

One basis for this oppositional stance may lie in the long-standing objection to institutions within some currents of Christian thought. As Weber remarked, "the separation of *charisma* from the person and its linkage with the ...*office*" is fundamental to the formation of the institutionalized church (Weber, 1978, p. 1148, emphasis in original). Originally, charisma (χάρισμα) meant "activities gifted by the Holy Spirit." To convey these through ministerial offices within the ecclesiastic hierarchy—bishops, priests, deacons—not only routinizes the gifts of the spirit but also administers the abundance of God's grace in institutionalized doses. For this reason, positing the sacraments as instruments of grace could be understood as continuing the centuries-long transformation of charismatic christian communities into juridical institutions, exemplified by the Catholic Church (Agamben, 2013, p. 9). A view of the sacraments as rituals that instrumentalize God's grace would align with Illich's understanding of the modern West as the perverse consequence of the effort to "institutionalize, legitimize, and manage Christian vocation" (Illich, in Cayley, 2005, p. 48). Therefore, whereas the soteriological effect of the sacraments would support a "religion of technology," the sacraments understood as instrumentalizing grace would support a religious objection to technology.

A second cluster of questions prompted by the essential nexus between theology and technology sketched above concerns the debate on the relative techno-scientific dominance of the European West over other civilizations (e.g., Chinese, Islamic, Indian). Around 900, western Europe was "hopelessly backward" in comparison to the Chinese, Islamic, and Indian societies. By 1914, Europe "owned" about 84 percent of the world (Hoffman, 2015, pp. 1–2). The literature on the reasons for this change in European fortunes is vast. The "Great Divergence" and the "Little Divergence" name the academic study of the disparity in technological development between the East and the West and within the West, respectively. Like many others of its kind, the academic debate on "the West and the Rest" (Ferguson, 2012) is polarized. One end of the pole argues for a continuous and cumulative growth in European techno-science that culminated in its decisive superiority, signaled by the Industrial Revolution in England. The other end opposes the very idea of "European" technology, arguing that the relative superiority of technical and economic development in the West was the result of transient and chance events.[4] However, to the best of my knowledge, all participants in this debate treat "technology" as an ahistorical category, as a natural kind. The debate on "the West and the rest" assumes both Su Song's water clock of 1094 in China and the horizontal axle windmill of 1185 in

[4] For a balanced review of this burgeoning literature, consult Mann (2017) and the introductory chapter of Davids (2013).

Yorkshire are technologies. By naturalizing artifacts, this debate begs the very question it seeks to answer.

Significantly, in the spirit of Max Weber, there is now widespread agreement in this literature that religious culture cannot be dismissed as a factor affecting economic growth and technological change in the West. In this regard, much ink has been spilled on the interpretation of Genesis 1.28 which, according to Lynn White's (1967) influential essay, authorized the domination of nature.[5] Contrarily, historians have convincingly shown that "rarely, if ever, did premodern Jews and Christians construe this verse as a license for the selfish exploitation of the environment" (Cohen, 1989, p. 5). Moreover, many biblical passages can be interpreted as encouraging a "caring" concern for creation. Hence, the idea that Christianity causes environmental destruction because it encourages dominating nature does not stand up to scrutiny.

However, less attention has been paid to the notion of salvation and its many articulations in Western culture as the fuel propelling techno-science. The idea of "salvation through technology" is perhaps not as laughable today as it was for Mumford over 70 years ago. Even, a cursory inspection of the advertisements for new devices, organizational arrangements, and social practices shows that belief in its this-worldly power to improve health, increase wealth, and foster happiness continues to animate technical innovations. It is therefore plausible that the original soteriological context of the instrument is not irrelevant to the continuing techno-science juggernaut that characterizes the West. What was once justified as conveying God's grace has been, over the centuries since Aquinas, defended by such ideas as the "relief of man's estate" (Francis Bacon), "convenience" (see Tierney, 1993), and "progress" (see Bury, 1920). Each of these tropes promises freedom from human travails through technology.[6]

Understanding the history of technology as instrumentalized salvation has a further implication. The rapid proliferation of devices in non-Western cultures may not be an expansion of "technology." The post-war boom in manufactured goods and services in Japan, Korea, China, and now India may not be accompanied by the perception of their distality and neutrality, and even less, by the promise of salvation. Both the faith in and fear of technology could be phenomena peculiar to the western provinces of the world.[7]

[5] Gen.1:28a: "Be fertile and increase, fill the earth and master it." Gen.1:28b: "Rule the fish of the sea, the birds of the sky, and all the living things that creep on the earth."

[6] I acknowledge the almost unforgivable short shrift paid to this point indicated by merely three citations. I also leave to this footnote the related point of the possible relationship between missions as a technology and saving the heathens since 1492.

[7] Yuk Hui (2016) defends his thesis that "in China, technics in the sense we understand it today —or at least as it is defined by certain European philosophers—never existed" (p. 10). While his argument is different from the one offered in this paper, the implications are similar.

8.5 Conclusion

The scientific study of "animal tool-use" serves as proof that, for the most part, technology continues to be imagined as an ahistorical phenomenon (Seed & Byrne, 2010). In contrast, in the 1990s, the late Ivan Illich argued that the category of the instrument is historical and that it emerged in the context of sacramental theology. His hypotheses have received confirmation in the recent works of two genealogists of biopolitics, Giorgio Agamben and Nicholas Heron.

Aristotle could not conceive the separation between the artisan and his tool because, for him, the latter was "part of" the former, as the body is part of the soul or as the slave is part of the master. For Aristotle, the potter was the efficient cause of the pot and no separate causality was ascribed to the potter's wheel. When referring to a minister of the Church, Thomas Aquinas used the notion of an "animate instrument" by which Aristotle defined the slave. Yet unlike Aristotle, Thomas divided efficient cause into a principal and an instrumental cause thereby giving the instrument a separate and distinct power. As an instrument, the minister has causative abilities—the ability to speak—which can be used in the service of God's work. Accordingly, the minister functions as an instrument of God used to indifferently transmit his saving grace. It is this two-fold aspect—distality and neutrality—that continues to inform our understanding of technology. By technology, we mean an object or convention designed with its own capacities—the car to move, the wheel to turn—which are then harnessed to the purposes of the user.

There are many more details to be filled in this account of the history of technology. Yet, the foregoing lends credibility to the argument that the relationship between theology and technology is internal and essential, not external and accidental. If that is correct, then a few consequences follow. First, the emphasis on the *salvational* effect of the sacraments could animate a "religion of technology," whereas the emphasis on the *instrumentalization* of grace could support a religious objection to technological innovation. Thus, both the religious encouragement of and the hesitance towards technology are explicable because of its origins. Second and relatedly, this two-faced dynamic could be relevant only to the West. Stated bluntly, the efflorescence of tools and devices in the post-WWII East need not be a technological phenomenon. Third, the relative eclipse of instrumental cause since the early modern period makes its rediscovery today something of a novelty. This rediscovery appears to have taken two, almost opposed, forms. On the one hand, it shows the historicity of technology, which implies both the beginning and the possible end of technology. On the other hand, the dominant currents in science and technology studies have discovered that things have agency, which could be little more than the unacknowledged rediscovery of Thomas' *agens instrumentalis* in a contemporary idiom.

Finally, technology as a category borne of theological reflections on the sacrament still carries a strong whiff of that soteriological promise. How else can one understand a paradoxical situation in the West today? For instance, it is widely accepted that anthropogenic emissions of greenhouse gases are the principal cause of the climate crisis. That is a roundabout way of saying that industrial society or techno-scientific civilization causes the climate crisis. And yet, the only acceptable solution to this problem seems to be more industry and techno-science, for example, in the form of electric cars and solar panels. Westerners continue to believe that technology can save them. The belief in "salvation through technology" stays unquestioned even as it fuels the burning of our world.

References

Agamben, G. (2013). *Opus Dei: An archaeology of duty*. Stanford University Press.

Agamben, G. (2016). *The use of bodies*. Stanford University Press.

Borgmann, A. (2022). Prospects for a theology of technology. In C. Mitcham, J. Grote, & L. Checketts (Eds.), *Theology and technology: Essays in Christian analysis* (Vol. 2, pp. 215–235). Wipf & Stock.

Bury, J. B. (1920). *The idea of progress; An inquiry into its origin and growth*. Macmillan.

Cayley, D. (2005). *The rivers north of the future: The testament of Ivan Illich*. Anansi Press.

Checketts, L. (2022). Conclusion. In C. Mitcham, J. Grote, & L. Checketts (Eds.), *Theology and technology: Essays in Christian analysis* (Vol. 1, pp. 132–141). Wipf & Stock.

Cohen, J. (1989). *Be fertile and increase, fill the Earth and master it: The ancient and medieval career of a biblical text*. Cornell University Press.

Davids, K. (2013). *Religion, technology, and the great and little divergences: China and Europe compared, c700–1800*. Brill.

Ferguson, N. (2012). *Civilization: The West and the rest*. Penguin Books.

Fudpucker, W. (2022). Through Christian technology to technological Christianity. In C. Mitcham, J. Grote, & L. Checketts (Eds.), *Theology and technology: Essays in Christian analysis* (Vol. 1, pp. 25–41). Wipf & Stock.

Heron, N. (2015). What is a minister?: Towards a theory of the instrumental cause. *The New Centennial Review, 15*(3), 135–165.

Heron, N. (2018). *Liturgical power: Between economic and political theology*. Fordham University Press.

Hoffman, P. (2015). *Why did Europe conquer the world?* University of Princeton Press.

Hui, Y. (2016). *The question concerning technology in China: An essay on cosmotechnics*. Urbanomic.

Illich, I. (1981). *Shadow work*. Marion Boyars.

Mann, M. (2017). Review article: The great divergence. *Millennium: Journal of International Studies, 46*(2), 1–8.

Mumford, L. (1975, March 10). Prologue to our time. *New Yorker Magazine*, 42–63.

Noble, D. (1999). *Religion of technology: The divinity of man and the spirit of invention*. Penguin Books.

Samuel., S. (2023). Understanding bureaucratic order: The theological paradigms of modern hierarchy. In G. Miller, H. M. Jeronimo, & Q. Zhu (Eds.), *Thinking through science and technology: Philosophy, religion, and politics in an engineered world* (pp. 255–275). Rowman & Littlefield.

Seed, A., & Byrne, R. (2010). Animal tool-use. *Current Biology, 20*(23), R1032–R1039.

Sismondo, S. (2010). *An introduction to science and technology studies* (2nd ed.). Wiley Blackwell.

Tierney, T. (1993). *The value of convenience: A genealogy of technical culture*. SUNY Press.

Weber, M. (1978). *Economy and society*. Guenther Roth and Claus Wittich (Eds.). University of California Press.

White, L. (1962). *Medieval technology and social change*. Oxford University Press.

White, L. (1967). The historical roots of our ecological crisis. *Science, 155*(3767), 1203–1207.

Chapter 9
Homo Deus or Homo Datum? Prolegomena to a Phenomenology of Dataism

Agostino Cera

Abstract This paper sketches a *Phenomenology of Dataism* ("the religion of data"), understood as the *transcendental technology* of our age, the barycentre of the digital world(s). Preceded by a *Laudatio* (a tribute to the figure and work of Carl Mitcham), a *Premise* (devoted to the epistemic crisis currently affecting the philosophy of technology), and an *Introduction* (which brings into focus the decisive transition "from Datafication to Dataism"), this phenomenology of Dataism is structured into *four parts*. First, an *Ontology of Dataism* (§ 1), which addresses the question: "What is Dataism?". Second, a *Genealogy of Dataism* (§ 2), which asks: "Where does Dataism come from?". Third, an *Anthropology of Dataism* (§ 3), which explores the question: "Where does Dataism flow?". The fourth and final part (§ 4) confronts the question: "What does Dataism essentially mean as a phenomenon?". The ultimate outcome of my argument is the *Protagora's Paradox*, according to which *pan-anthropism* itself (i.e., the entelechy of the anthropocentric *Weltanschauung*) requires the de-humanization of the human being—or, more precisely, the "disenchantment of the world" presupposes the "self-disenchantment of the human being". *Homo deus* thus emerges as a mask behind which *homo datum* is concealed: a believer in *techne deus*, that is, an adherent of the religion of technology.

Keywords Dataism · Philosophy of technology · Anders (Günther) · Harari (Yuvàl Noah) · Pan-anthropism

A. Cera (✉)
Università di Ferrara, Ferrara, Italy
e-mail: agostino.cera@unife.it

L. Checketts (ed.), *What Went Wrong with the West: Carl Mitcham's Critique of Western Philosophy of Technology*, Philosophy of Engineering and Technology 52, https://doi.org/10.1007/978-3-032-22214-5_9

9.1 Dedication/Laudatio. To "a (Philosophical) Hero of the Two Worlds"

Despite the vast and burdensome past behind us (often even before our very eyes), we Italians are a young people and nation. "*L'unità d'Italia* (the unification of Italy)" was, in fact, only achieved between 1861 and 1871. Italy often defines—and recognizes—itself as the "*paese dei campanili* (country of bell towers)", because it is characterized by tiny homelands often pitted against one another, chronically incapable of standing together. The difficulty in feeling Italians is well expressed by a phrase attributed to Massimo D'Azeglio (1798–1866), one of the political protagonists of the unification process. He is said to have remarked: "*fatta l'Italia, bisogna fare gli italiani* (we have made Italy; now we must make Italians)". This explains why our national pantheon is not very populated. Among the few symbols with which Italians can identify, there is certainly the practical—i.e., military—architect of the unification process (while the political one was Camillo Benso, Count of Cavour): Giuseppe Garibaldi (1807–1882), perhaps the only real "Italian national hero". Garibaldi was known as "*l'eroe dei due mondi* (the hero of the two worlds)" because of his military exploits not only in Italy and Europe, but also in South America.

This strange prelude, seemingly entirely out of place, is perfect to introduce this premise, which is actually a dedication; or, even better, a *laudatio*. As an Italian, indeed, thinking about Carl Mitcham and his long and prestigious career as a scholar in the field of the philosophy of technology and STS studies, the figure of Garibaldi often came to my mind. In my eyes, Mitcham represents a kind of *philosophical hero of the two worlds. And of the two cultures.* By this definition, I refer to the fact that, among his many qualities, the one that, in my opinion, best identifies him is *ecumenism*, born of an extraordinary intellectual curiosity and an uncommon "*Goût des autres* (taste of other)".[1] These are rare virtues in general, not only in the academic world.

In addition to recognizing, cultivating, and safeguarding the *indispensability of the historical dimension for the philosophy of technology*[2]—in contrast to the presentist inertia, typical of the current mainstream in this field –, Mitcham early on understood the need to overcome traditional disciplinary boundaries, starting with the fundamental divide between "the two cultures" (Snow, 2012). Without considering his pioneering work in the 1970s with Robert Mackey (Mitcham & Mackey, 1972, 1973), which would lead, in 1975, to the first conference meeting of the Society for Philosophy and Technology (SPT)[3] organized by Paul Durbin at the University of Delaware, Mitcham's most famous book, *Thinking through*

[1] See (Jaoui, 2000).

[2] Hans Achterhuis called him "the most important historian of the philosophy of technology" (Achterhuis, 2001, 4).

[3] Mitcham was the first SPT president, from 1981 to 1983.

Technology: The Path between Engineering and Philosophy (Mitcham, 1994),[4] now a classic, embodies both these inspirations. On one hand, it represents a valuable reconstruction of the historically crucial stages in the philosophy of technology, and on the other the building of a bridge for dialogue between the extreme fronts of the two cultures: philosophy and engineering. Such an ecumenical approach was made necessary by the peculiarity of modern technology understood as an epochal phenomenon, that is, as the current "subject of history" (Anders, 2013, 9).

However, unlike several prominent figures of the so-called empirical turn, Mitcham has always based his commitment on the recognition and defense of *philosophy's epistemic difference*: its uniqueness and irreplaceability as a form of knowledge, its irreducibility to a "positive science." His entire body of work testifies to a genuine faith in the *philosophy* of technology. This is the essence of what I like to call *co*-disciplinarity: an authentic relationship between different epistemic regimes, starting from the acknowledgment and respect of the other's diversity, namely from the fact that the other is—and must remain—"different from me." This kind of attitude goes beyond the best-known concepts of *inter*-, *multi*- and *trans*-disciplinarity.[5]

Earlier, I spoke of "intellectual curiosity" and "*Goût des autres*" as defining traits of Mictham's personality. As further evidence of this, let me highlight his long-standing attention to authors and movements, both past and present, that operate outside the academic and publishing mainstream. Consider, for instance, his interest in two "heretical" philosophers of technology like Jacques Ellul and Ivan Illich,[6] or in fascinating but decidedly isolated projects like Ernesto Mayz Vallenilla's "meta-technics" (Mayz Vallenilla, 2004), or in the whole tradition of philosophy of technology in the Spanish-speaking world (Mitcham, 1993). Before its definitive worldwide success, Mitcham also showed a strong appreciation for the project of an intercultural philosophy of technology, that is, a "cosmo-technics," carried forward by Yuk Hui.[7] Based on all these reasons, Mitcham's approach should be considered a *polyphonic ecumenism*, within which even the less heard voices must find space, provided they prove to be philosophically worthy of interest.

Another word/virtue that aptly characterizes Mitcham's human and intellectual figure is undoubtedly *generosity*. In this regard, let me draw from my personal experience. Nearly 10 years ago, as a complete unknown (to this day I am far from being "known", but at the time I had only published in Italy and in Italian), I attempted to contact "Prof. Mitcham" in person—i.e., one of my reference scholars—presenting him with a summary (in English) of my work. I had little hope of receiving a response, which instead came promptly and was immediately an offer for an

[4] Testifying to a 30-year interest in this topic, see also (Mitcham, 2020).

[5] Mitcham was co-editor of *The Oxford Handbook of Interdisciplinarity* (Frodeman et al., 2010) and editor in chief of the *Encyclopedia of Science, Technology and Ethics* (Mitcham, 2005).

[6] See, respectively, (Jerónimo et al., 2013) and (Hoinacki & Mitcham, 2002).

[7] See his review of Yuk Hui's *The Question Concerning Technology in China: An Essay in Cosmotechnics* (Mitcham, 2018) and, more recently, of *Post-Europe* (Mitcham, 2025).

authentic/equal dialogue—because only a dialog of equals is authentic dialogue—despite the vast gulf of prestige and authority that separated, and still separates, us.

In 2023, with the same generosity, Carl agreed to participate in a round table on my book on the Anthropocene (Cera, 2023), held during the SPT Biennial Conference in Tokyo. What struck me most during those Japanese days, was seeing Carl "jumping" from floor to floor of the building hosting the conference: not just for his own commitments, but especially to accompany and support his young Chinese students in their presentations. As is well known, after many years at the Colorado School of Mines, where he is Emeritus Professor, he has more recently become Visiting International Professor of Philosophy of Technology at Renmin University of China (indeed, in "another world"), where he is nurturing new generations of scholars. During a break in the conference, wondered by what I had noticed, I approached him to tell him that his enthusiasm was contagious and to ask him how he managed—I wanted to add: "at over 80 years of age!"—to have all that energy. With the greatest naturalness, Carl replied: "I love teaching."

This, then, is what I hope to have learned from scholars like Carl Mitcham. Like another master in the field of the philosophy of technology who recently passed away, Albert Borgmann, Carl is one of those rare examples of a scholar who reminds us that philosophy is, above all, a way of life. Or better, a way of being-in-the-world; even in "two worlds."

With sincere gratitude for what he has taught us and will continue to teach us, I allow myself to dedicate the following paper to him, which deals with the topic of dataism. Despite its inspiration being not ecumenical enough (too critical), I really hope he will appreciate it.

9.2 Premise: An Epistemic Crisis

The basic assumption of this paper is the same as that of a work I have been doing for several years now, namely, that the philosophy of technology is living an *epistemic crisis*. More precisely, an involution from the so-called empirical turn to what I have called *ontophobic turn*.[8] With this expression I mean that some of the most prominent scholars in this field *de facto* don't believe anymore in both "philosophy" (as a peculiar form of knowledge) and "Technology" (as something in itself, as an epochal phenomenon). To avoid being dystopian or apocalyptic—as partly the classical philosophy of technology was—the current mainstream in this field has become *apologetic*, that is, no more able to be critical. It is flattened on the status quo, incapable to distance itself from the reality. In front of the real—of what it is—it can say only "Yes and Amen" (Nietzsche, 2006, 184–87). Listening to the scholars referable to this attitude, you have the impression that we are really living in "the best of all possible worlds." This epistemic crisis, which equates to an *apologetic shift* or *obstruction of critique*, manifests itself concretely in the metamorpho-

[8] On this topic see (Cera, 2020).

sis of the philosophy of technology into a *problem-solving activity*, that is, a search for solutions in front of the concrete problems emerging from the single technologies. Probably the most glaring example of this *solutionist drift* of the philosophy of technology can be found in the plethora of literature devoted to self-driving vehicles, within which moral dilemmas turn into unintended advice for insurance companies.

With reference to this whole state of things, my objection is the following. If Technology with a capital 'T' *becomes nothing*, namely if in itself it means nothing, but at most the umbrella term of the single technologies, then the paradoxical but consequential result of this situation is that the philosophy of technology ceases to have any meaning/value in itself. In other words, the paradoxical accomplishment of the empirical turn as an ontophobic turn should be the *self-suppression*, or at least *self-overcoming*, of the *philosophy* of technology.

Given this assumption, my point is that the time has come for a reflection on the epistemic status of the philosophy of technology (even of philosophy itself). By presenting a critical reading of an epiphenomenon—i.e., dataism—of Technology understood as an *epochal (Ur-)phenomenon*, or an "integral epochal phenomenon", the following pages are intended as a contribute to stimulate such a reflection.

9.3 Introduction: From Datafication to Dataism

In her 2014 paper, José van Dijck defines *datafication* "as a new paradigm in science and society" (2014, 198). In doing so, she refers to the book of Mayer-Schoenberger and Cukier, according to which datafication is "the transformation of social action into online quantified data" (2013, 30). More than 10 years after these investigations, it can be argued that this was still a partial definition of a phenomenon that has since shown far greater scope and ambitions. Today, in fact, datafication is proving to be the attempt to transform every entity, without exception, into (online) quantified data. Insofar as it is characterized as a project of *universal ontological conversion*—a comprehensive reconfiguration of reality –, datafication becomes *dataism*: no longer a scientific or social paradigm, but an *ideology*, even a *worldview*. The world shaped by datafication—the "datafied cosmos"—equates to the so-called "infosphere."[9] More precisely, the infosphere is the form of the world expressed by the dataism as a *Weltanschauung*/ideology of datafication.[10]

[9] According to Luciano Floridi, the infosphere is "the world of data, information, knowledge and communication" treated "as a new environment" The infosphere "is an environment that is essentially intangible and immaterial but not, for this reason, any less real or vital" (Floridi, 2001). See also (Floridi, 2014, 25–58), where, among other things, it is possible to read this emblematic Hegelian paraphrase: "what is real is informational, and what is informational is real" (Floridi 2014, 41).

[10] Kobelieva and Nikolaienko argue that "with the advent of Big Data processing algorithms, a new ideological paradigm appears—dataism, which predicts the merger of a person with the general

As the crystallization (hypostatization) of the datafication process, dataism embodies one of the most advanced fronts of *technolatry*: the idolatry of technology, an investment not only of trust but of faith in the possibilities that technological development offers us. A soteriological aspiration towards it. Technolatry is the real face of the current human-technology "relation." Here and now, in fact, we ask from technology exactly what we have traditionally asked from religious forces: to be saved, that is, to be redeemed from the condemnation of our finitude. With respect to a phenomenon of this magnitude, on an analytical level it is not enough to observe that datafication affects the worldview or that it is turning into a worldview itself (as a "data worldview").[11] That is to say, it is not enough to limit one such observation, without trying to investigate the unexpressed assumptions of this phenomenon; precisely, its ideological dimension. In my opinion, the limits of those investigations, which were nonetheless able to grasp some crucial features of datafication,[12] lies in their *unintentional apologetic temptation*: their inability to move from observation (statement) to judgment (criticism).

From an epistemic perspective, such an attitude could be justified by an appeal to that "*Wertfreiheit* (value freedom)" described by Max Weber as the scientific *habitus par excellence*.[13] However, what we are realizing is that, in the long run, this "axiological neutrality" does not remain neutral but takes a position in terms of apologetics and justificationism, namely as an *obstruction of critique*. Gradually, that *passive neutrality* turns into an *active neutralization*, a systematic naturalization of everything it encounters. As a result, this *obstruction* of critique turns into a *castration* of critique, a true inability to produce any judgment. The main outcomes of this critical inhibition can be seen and measured on the ethical and political levels. Regarding the "question concerning technology", i.e. technology as an epochal phenomenon, this kind of approach means legitimizing any concrete manifestation of it, in the sense of never finding objections of principle against it. And this is exactly the logic of problem-solving, since "problem"—that is, the transformation of any question into a problem—essentially means "the promise of a solution", even when, in practice, it cannot be obtained.[14]

Having said that, it should also be added that this chronic lack of critique in philosophy does not leave a desert, an empty space, but rather something that

data flow". Despite this assumption, their analysis bets on the normalization/naturalization of this "process", which is considered "a completely logical continuation of human evolution" (Kobelieva & Nikolaienko, 2021, 100).

[11] (Romele, 2023).

[12] in addition to the aforementioned (Romele, 2023), this is the case with Orit Halpern's albeit interesting work. See (Halpern, 2014) and (Halpern & Mitchell, 2023).

[13] On this topic see (Weber, 2012a, 304–34).

[14] This is the spirit of what Evgeny Morozov has called *solutionism*, that is, "the belief"—the faith, I would say—"that every problem has a solution based in technology" (Morozov, 2013, 1–16). According to Jacques Ellul, the coming of the rise of the "technological system" takes place when technology's *ratio operandi* creeps in any human context, that is, when "in every field men seek to find the most efficient method" (Ellul, 1964, 21).

technology (i.e., its logic, its *modus operandi*) goes on to occupy - to sature, to be more precise Consequently, the new imperative of our time as the age of technology is the so-called "Gabor's law", which reads: "*what can be made, must be made* [...] what must be made is inevitable" (Anders, 2013, 17), or "all that can be done must be done" (Ellul, 1980, 207). What is technically possible not only becomes permissible but acquires a normative character, transforming into a moral necessity. Given this unprecedented epochal framework, the more technology impacts the way we perceive and experience reality, to the point of reshaping and even replacing it (thus becoming a "neo-environment"),[15] the more philosophical reflection should take on the responsibility of judgment and possible critique.[16] As always, indeed, an ideology is not a simple description of the world/reality but an interpretation of it, even an attempt to (re)shape it. In this specific case, this means grasping the *hidden normative nature of dataism* (i.e., as the ideological evolution of datafication), which is expressed in the belief/faith that every entity can be reduced to quantifiable data, that everything escaping this conversion is to be considered ontologically devoid of value. In a formula, which we will delve into shortly: "*what is not—or cannot become—datum, is not at all.*"

Moving from these assumptions, the following pages represent an attempt to bridge this gap between experience and judgment, to rehabilitate the critical dimension of philosophical work. As anticipated, I will sketch a *phenomenology of Dataism*: "the religion of data", the most faithful characterization of our current perception of reality. My basic claim is that this neo-religion stands for the *transcendental technology*—i.e., the paradigmatic form of technology—in our age. More precisely, dataism represents the keystone of the digital turn or the barycenter of the digital world(s).

This (proto-)phenomenology of dataism consists of the following parts:

1. an *Ontology* of Dataism, that is, its general definition and characterization. It tries to answer the question: *What is* Dataism?
2. a *Genealogy* of Dataism, that is, an overview of its historical-theoretical background. It tries to answer the question: *Where is* Dataism *from?*
3. an *Anthropology* of Dataism, that is, the description of human type secreted (produced) by this neo-environment. It tries to answer the question: *Where* Dataism *flows?*
4. the *Conclusions*, which try to answer the question: *What essentially* Dataism *means* as a *phenomenon?*

[15] On this topic and the related establishment of a "*Philosophy of Technology in the Nominative Case* (TECNOM)", see (Cera, 2017).

[16] As is known, "judgement" and "critique" are synonymous in Greek *krinein.*

9.4 An Ontology of Dataism

The most neutral interpretation of dataism is its characterization as an evolution of datafication process. More precisely, once the latter proves to be a project of *universal ontological conversion*, then *dataism* takes place. As a result, it is no longer a scientific or social paradigm, but an *ideology*, an authentic *worldview*. The world shaped by datafication—the "datafied cosmos"—equates to the so-called "*infosphere*" as the outcome of a "fourth revolution" and "third-order technology", according to Luciano Floridi's lexicon (see Floridi, 2014).

Dataism establishes the advent of the *Big Data Age*, that is, it emphasizes the matter of fact that the big data phenomenon gives birth to its own story. This *story* gradually becomes a *narrative*, then a *point of view*, then an *ideology*, then a *worldview* and finally a kind of *religion*, in the sense of that particular kind of storytelling which asks faith and promises salvation. In a formula: where big data narrative goes from being a story to being a religion (i.e., a techno-religion), there dataism happens.

While the natural father of dataism should be considered the journalist David Brooks—who used this word (in the form "Data-ism") within an article about *The Philosophy of Data*, published on *The New York Times* in February 2013 (Brooks, 2013)[17]—and while the first book dealing with this topic was published by the journalist Steve Lohr in 2015 (Lohr, 2015), the putative father of this neologism—he who made it a global trend topic—is without any doubts Yuvàl Noah Harari, in his 2015 bestseller *Homo Deus*. According to Harari, dataism embodies the most concrete form taken by the "techno-religion" as a prevailing cult of our age. The only still possible cult of our age. The irreversible crisis of humanism (also in the form of "techno-humanism") brings out alternative solutions, among which emerges a techno-religion that "seeks to sever the humanist umbilical cord altogether" (Harari, 2015, 366). Technology stands out therefore as the only *idolum* surviving the "death of God" or the unique divinity still plausible (believable) in the age of the accomplished nihilism. Such a techno-religion represents the counterpart of a corresponding state of mind/feeling: *techno-latry*, that is, human being's worship/idolatry of technology.

Harari asks himself: "What might replace desires and experiences as the source of all meaning and authority? As of 2016, only one candidate is sitting in history's reception room waiting for the job interview. This candidate is information. The most interesting emerging religion is Dataism, which venerates neither gods nor man—it *worships data*" (Harari, 2015, 366—my italics).

In Harari's view dataism, understood as the vanguard of techno-religion, "says that the universe consists of data flows, and the value of any phenomenon or entity is determined by its contribution to data processing" (Harari, 2015, 367). As a result, dataism's final purpose can be found in a project of universal ontological conversion

[17] Brooks defines data-ism as "the rising philosophy of the day", which consists in the fact that "we now have the ability to gather huge amounts of data" (Brooks, 2013).

as a comprehensive reconfiguration of reality; a "reontologization", according to Floridi - or a real *transubstantiation*, according to a religious-theological lexicon.

In observance of the techno-religious *forma mentis*, any new technology, as an *emerging* technology, aims to be a *disruptive* technology.[18] This means that within such an ideological framework the *disruptiveness* stands out as a necessary feature of any technology, the reference parameter to evaluate it. A single technology can be considered a "good" technology (i.e., successful, efficacious…), precisely insofar as it proves to be a disruptive technology. In turn, this *compulsion to disruptiveness* as *telos* of any single technology equates to the best guarantee for the "*total mobilization*"—the hysterical dynamism of an endless and purposeless iteration—as a prevailing paradigm of the present reality.[19] A disruptive technology, indeed, represents the best guarantee to produce a shock—or a "creative destruction", according to the famous expression by Schumpeter—to the reality, namely to avoid its standing still (i.e., stagnation), because the standing/stagnation counts as the only mortal sin still valid in the brand-new Eden promised by techno-religion. So, *disruptiveness* means *shock*, shock means *movement/mobilization*, movement/mobilization means *no standing still/static*. A well-known Italian expression sums up well the spirit of this paradigm. It says, "*chi si ferma è perduto* (those who stop are lost)." As *telos* of any single technology, disruptiveness embodies the utopia/myth of a permanent/ uninterrupted "change" of the *status quo*. It establishes the *flow* as an epochal imperative and thus confirms the essentially *liquid* character (in the sense of Zygmunt Bauman) of the current "network society".[20]

Having said that, one could argue that, ontologically speaking, we are faced here with an updated version of the gigantomachy between "*stasis* (rest)" and "*kinesis* (motion)", namely two of the five "*megista ton genon* (greatest of the kinds)," presented in Plato's *Sophist*.[21] The difference is that after a couple of millennia the hierarchy between these two kinds is totally inverted. While Plato's ontology considered *stasis* something inherent to *eidos* and thus "*agathon* (the good)", now this positive pole (*agathon*) corresponds to *kinesis*. On the contrary, now *stasis* means "*kakon* (the evil, the defect)". The traditional *horror vacui* has now turned into *horror stasis*. Translated in Nietzsche's antiplatonic words, the techno-Eden—the "data hyperuranium", so to speak—equates to a configuration of reality where the dream "to impress upon *becoming* the character of *being*" comes finally true. Incidentally, it is worth remembering that Nietzsche considered this dream as "the highest expression of the will to power" (Nietzsche, 2017, 353—fragment 617).

---

[18] According to the *Cambridge Business English Dictionary*, a disruptive technology is "a new technology that completely changes the way things are done" (Cambridge Dictionary, 2020). The reference book on this topic is (Christensen, 1997).

[19] Obviously, I refer here to the well-known formula coined by Ernst Jünger in his 1930 essay (Jünger, 1998).

[20] According to Manuel Castells, the *Network Society* is "the social structure that results from the interaction between social organization, social change, and a *technological paradigm* constituted around digital information and communication technologies." (Castells, 2004, xvii).

[21] See (Plato, 1921, 402–13; 254b-256d).

Dataism, as said, stands as the peak of this whole process: the uninterrupted and unstoppable "data flow" is the perfect implementation of the total mobilization as an epochal imperative, the best mirror for the new hierarchy among the "greatest of the kinds". As a *worship of data* (and devotion to their flow), dataism emerges when emerges the firm belief—the faith—that datum is *the ultimate configuration of all entities*, their standard ontological form. If we look at dataism as a language, then datum represents its minimum unit (syntagm), while algorithm represents its phrases or sentences and algorithm's rules of combination its grammar or syntax. According to a religious metaphor, instead, if dataism embodies the orthodoxy of techno-religion (the prominent manifestation of technolatry), then datum embodies its fundamental dogma and its basic form of incarnation,[22] while the algorithm equates to its sacraments or liturgy.

Within such a framework, datum emerges as *the current: logical, epistemic, alethic and finally ontological standard*. That is to say, something can: make sense, be true or false, and exist, only insofar as it is—or can become—a datum. As a consequence, all that is not—or cannot become—a datum equates to illusion, mistake and falsity. It has no ontological legitimacy. *What is not—or cannot become—datum, is not at all*. By paraphrasing Giambattista Vico's well-known motto, one could argue that here and now "*verum et datum convertuntur* (the true and the datum are convertible)."

Dataism's basic assumption can be found in the *ontological equation between being and being computable* and its potential epigraph would sound: "*esse est computari*" (see Capurro, 2010), an updated version of the classical Berkeley's "*esse est percipi*". This equation highlights that a datum establishes, by definition, the *total reduction of an entity to its computability*, its transubstantiation as a complete dematerialization in the form of digitalization. In turn, this digital dematerialization (which, incidentally, reveals a kind of gnostic aftertaste in its disregard for the "flesh")[23] is functional to a project of dominion and subjugation, because this *techno-catharsis* seeks to realize the utopia of a completely manageable reality. Indeed, once an entity is entirely turned into a set of data (i.e., into something computable), it becomes completely forseeable, controllable and manageable; something entirely available to us, at our complete disposal. By means of dataism finds expression a *will to power 2.0*, according to which the natural *pendant* of the total mobilization is the "*total administration*" (Marcuse, 2002, 88ff., 107ff.), namely the establishment of a "*planetary management*."[24]

Following its ideological/religious function and meaning, dataism can be seen as the fulfilment of Galileo's sentence which summarizes the spirit of modern science, and modernity itself more in general. Galileo stated that "the book of nature is

[22] Incidentally, here would take place the oxymoron of a *digital incarnation*, that is, the paradox of an individuation without matter, disembodied.

[23] On this topic, as further evidence in favor of techno-religion's idea, see (Davis, 1998; Noble, 1999; Campagna, 2018).

[24] This formula gives expression to Ecomodernist post-environmental utopia: the ecological version of techno-latry. On this topic see the *Ecomodernist Manifesto* (Asafou-Adjaye et al., 2015).

written in mathematical language and the symbols are triangles, circles and other geometrical figures, without whose help it is impossible to comprehend a single word of it" (Galilei, 1960, 184). Insofar as dataism reduces everything to its computability, it carries to the extreme this assumption by stating that "the book of nature, which is written in mathematical language, is itself (i.e., as an object/entity) nothing but a set of mathematical characters (or bit). The book of nature is (can be made) an ebook". Or, if you prefer, "it form bit", according to the dogma of pan-computationalism established by John Achibald Wheeler (2018). According to Luciano Floridi: "Galileo thought nature was like a book written in mathematical symbols that could be read by science. This might have been a metaphorical stretch at the time, but today the world we live in is certainly becoming more and more a book written in digits" (Floridi, 2023, 204).

9.5 A Genealogy of Dataism

Given its *modern imprinting*, from a historical-philosophical perspective dataism can be held as the final step of that "disenchantment of the world" which Max Weber saw as the main goal of modernity. In his well-known conference *Science as a Vocation* (1917), he defines the "disenchantment of the world (*Entzauberung der Welt*)" as "the increased intellectualization and rationalization", that is, "the knowledge, or the belief, that *if we wished to*, we *could* at any time learn about [...] that we can – in principle – *dominate* everything by means of *calculation*." (Weber, 2012b, 342).

Such a process strives to realize *homo faber*'s (*homo technologicus'*) utopia: the maximum extent of our will to power, the building of a totally anthropized world, a configuration of reality definitively human-scaled and anthropo-centred, entirely subjugated to our *desiderata*. After more than 2000 years, Protagoras' claim for *metron anthropos* ("man is the measure of all thing"—DK 80B1) can finally come true. I call this phenomenon *Pan-anthropism*, which represents the apotheosis—or *entelechy*—of anthropocentrism, the advent of *homo deus* as the natural evolution of *Übermensch*. According to Harari, "having raised humanity above the beastly level of survival struggles, we will now aim to upgrade humans into gods, and turn *Homo sapiens* into *Homo deus*" (Harari, 2015, 35).

The following words by Werner Heisenberg—pronounced, more than half a century ago, to characterize "the idea of nature in contemporary physics"—gives us a perfect "inner portrait" of the pan-anthropic scenario: "for the first time in the course of history modern man [*homo faber*] on this earth now confronts himself alone." (Heisenberg, 1958, 23). The coming of dataism marks therefore the establishment of a completely disenchanted world, a fully realized pan-anthropism, and the evolution from *Übermensch* to *homo deus*.

Concretely, we can find at least *three stages* in this process, that is, three different interpretations of both world and human being, and their relation three ways of disenchantment that gradually lead to the pan-anthropic goal.

1. The first stage is the *Cartesian Stage*. Here "entity" in general is interpreted as *objectum* (object): something which *is in contrast to* human being as *subjectum* (subject). In this case the reality is interpreted as an *obstacle* to human agency, an obstruction to the achievement of its ends. As a result, the relationship between human being and world takes the form of a *struggle*. The epigraph of this first stage could be this "*à la* Descartes" motto: "*mundus est machina perfecta* (the world is a perfect machine)".
2. The second stage is the *Vichian Stage*. Here "entity" is interpreted as *factum* (fact): something which *stands in front of* human being as maker, that is, *homo faber*. In this case the reality is interpreted as a *product* of human agency, a pure means to the achievement of its ends. As a result, the relationship between human being and world takes the form of a *production relation*. The epigraph of this second stage could be Vico's motto: "*verum et factum convertuntur* (the true and the made are… convertible)" (Vico, 1971, 65).
3. The third stage is the *Heideggerian Stage*. Here "entity" is interpreted as *Bestand* (standing-reserve)[25]: something which *is subjugated to* human being as *creator*, that is, *homo creator*. In this case the reality is interpreted as a mere *projection* of human agency, a pure occasion/pretext to the achievement of its ends. As a result, the relationship between human being and world takes the form of what Heidegger calls "*challenging* (*Herausfordern*)".[26] The epigraph of this third stage could be his famous statement: "the essence of technology is by no means anything technological." (Heidegger, 1977, 4) or Melvin Kranzberg's first law of technology, according to which "technology is neither good nor bad, nor is it neutral." (Kranzberg, 1985, 50).
4. More than a fourth stage of this process, dataism should be considered a *sub-stage* within the Heideggerian stage, that is, a variaton on the canon established by the interpretation of entities as standing-reserve, or better, its definitive accomplishment. The interpretation of "entities" as *data* corresponds to the fulfilled disenchantment of the world, its total reduction to human measure. Indeed, as a datum—i.e., something entirely computable—an entity loses even its material consistency, that is, the basic form/last trace of its autonomy and otherness. As a datum, it becomes once and for all something totally at our disposal, in our complete availability. Dataism, as said, establishes the triumph of pan-anthropism: a scenario within which the relationship between human being (as *homo deus*) and world (as "digital *Bestand*", so to speak) takes the form of total management and control. Under the commandment of the *compulsion to disruptiveness*, under the "principle of perpetual shock" and "the standing still taboo", the reality turns into the space of both "total mobilization" and "total

[25] "The name 'standing-reserve (*Bestand*)' […] designates nothing less than the way in which everything presences that is wrought upon by the challenging revealing. Whatever stands by in the sense of standing-reserve no longer stands over against us as object" (Heidegger, 1977, 17).

[26] "the revealing that rules in modern technology is a challenging (*Herausfordern*), which puts to nature the unreasonable demand to supply energy that can be extracted and stored as such" (Heidegger, 1977, 14).

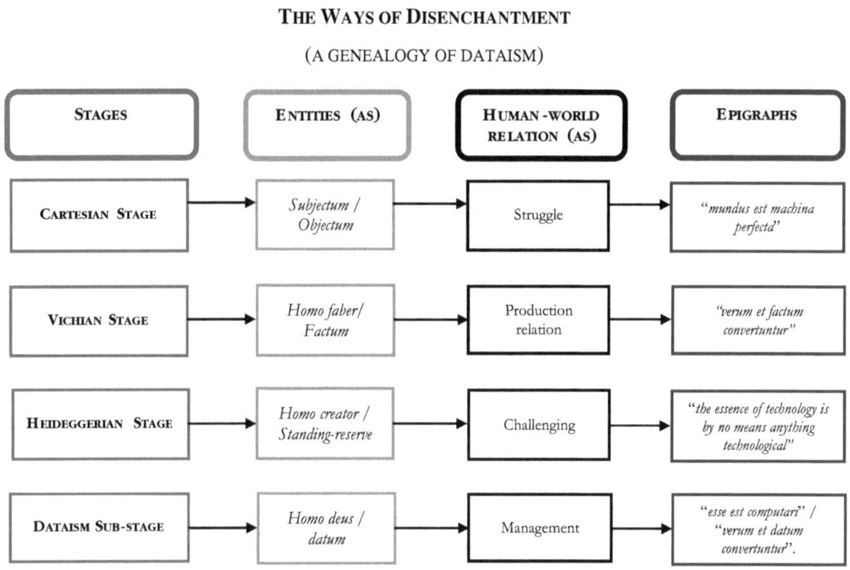

Fig. 9.1 The ways of disenchantment (A genealogy of dataism)

administration".[27] The epigraph of this final sub-stage could be the already quoted mottos: "*esse est computari*" or "*verum et datum convertuntur*" (Fig. 9.1).

9.6 An Anthropology of Dataism

However, beyond its apparently granitic consistency, at the basis of dataism as a worldview/ideology and new configuration of the reality hides a crack, that is, a paradox. An *anthropological paradox*, to be more precise, which in turn reveals the internal aporia of disenchantment of the world, pan-anthropism's inherent contradiction.

To adequately highlight this paradox, I will refer to Günther Anders' "philosophical anthropology in the epoch of technocracy" (Anders, 2013, 9), an investigation of our age in many ways still unsurpassed. The basic assumption of this approach is the acknowledgement—as an epochal matter of fact—that technology represents the current "subject of history" (Anders, 2013, 9). According to Anders, "we have renounced […] considering ourselves as the subjects of history; we have abdicated, replaced ourselves with another subject of history: technology, the history of which has become *the* history over the course of recent history." (Anders,

[27] As said, the total mobilization represents the condition of possibility for the total administration, i.e. for an entirely manageable (entirely disenchanted) world.

2013, 279). Such an abdication corresponds to the "proletarianization" of human being, its downgrading to a mere "co-historical" role.[28]

Anders also defines his thought "discrepancy philosophy", since it focuses on the "ever increasing discrepancy between what we can produce and what we can imagine" (Anders, 1985), that is, on our imagination's growing incapacity to keep up with the frantic pace of our technological capability. At the heart of his hermeneutic of the present, he places a *double anthropological metamorphosis*, by means of which human being tries to make true its pan-anthropic dream, to totalize its will to power by extending it on every entity. It is the metamorphosis of *homo faber* into *homo creator* and of *homo creator* into *homo materia*.

1. The triumph of technology as a planetary force transforms *homo faber* into *homo creator* (*homo deus*, according to Harari) . *Homo creator* represents the human type produced by the evolution of *techne* from *mimesis* (imitation) to *poiesis* (creation), a transition that takes place when technology becomes able not only to produce, but rather to create, that is, when "*physis* is produced by means of *techne*" (Anders, 2013, 21). The *homo creator condition* corresponds to the situation in which *homo faber* absolutizes his Prometheanism, by extending it to the totality of entities without exception. As *homo creator/homo deus*, human being feels like and acts as the lord of a world/nature interpreted and made use of as an object or standing-reserve. Here the entelechy of anthropocentrism takes place; here the pan-anthropic dream comes real. Otherwise, this realized dream implies a cost. A hidden cost. As said, in fact, in order to become a *creator/deus*, *homo faber* must transform *everything* into raw material or *Bestand*. "Everything", therefore also himself.
2. Based on this premise, the paradoxical but consequential outcome of the metamorphosis from *homo faber* to *homo creator/homo deus* is its complementary metamorphosis into *homo materia*. To be the authentic subject of present reality—"the *kybernetes* (helmsman) of total mobilization", or "the CEO of total administration"—human being must make itself the object of its own making. It must subjugate itself to its own technological omni-power, thus becoming an object among objects, a standing-reserve among standing-reserves. By doing so, technology acquires a universal/epochal status, thus becoming the "subject of history": Anders' characterization of the same historical-ontological phenomenon that Heidegger calls "*Ge-stell* (Enframing)".[29]

If everything is measured, employed and evaluated by its "challengeability", namely if everything exists insofar as it is (or can become) *Bestand*, raw material or artifact, then the human being obeys this epochal injunction by becoming a *human raw material*, that is, a *Bestand-Mensch* (standing-reserve human being), or, to use a

[28] On the topics "proletarianization" and "co-historicity", see (Anders, 2013, 297).

[29] "Enframing means the gathering together of that setting-upon which sets upon man, i.e., challenges him forth, to reveal the real, in the mode of ordering, as standing-reserve. Enframing means that way of revealing which holds sway in the essence of modern technology and which is itself nothing technological" (Heidegger, 1977, 19).

more familiar expression, a *human resource.* In Anders' view, this anthropological paradox equates *ipso facto* to a moral paradox: in two "supraliminal"[30] events as Auschwitz and human engineering projects he sees a concrete evidence of *homo materia*'s emergence.

9.7 Homo Deus or Techne Deus?

We are now able to draw conclusions of this (proto-)phenomenology of dataism. As we have seen, to make true the pan-anthropic utopia of a totally manageable/computable world (i.e., the goal of the whole modernity), human being—as *homo faber*—must sacrifice itself to the technological *idolum.* It must transform technology into an object of faith by charging it with a religious meaning and a saving value. According to Harari's vocabulary, it must establish a techno-religion. This is what I mean with "*techno-latry*".

Well, the inherent contradiction of this project—we can call it *pan-anthropic aporia* or *Protagoras's paradox*—depends on the fact that *pan-anthropism itself demands the de-humanization of human being*, that is, its metamorphosis into *homo materia.* As a result, the disenchantment of the world necessarily requires the *self-disenchantment of human being*, (a "human apoptosis", to paraphrase Floridi's lexicon),[31] which in turn consists in the establishment of a techno-religion, namely a profession/confession of techno-latry.

The absurd, but entirely consequent, introjection of this *technolatric dispositive*, according to which we let ourselves be: enhanced, corrected, healed and finally saved from that which we ourselves produced (from our technological omni-power), is what Anders calls "*Promethean shame*". It is that shame, "which one feels when faced with the 'humiliatingly' high quality of the objects we ourselves make" (Anders, 2002, 23), or better, the outcome of a "*Promethean gap*" created by "the inability of our soul to be 'up to date' with our production" (Anders, 2002, 16).

To cross the threshold of the techno-Eden, to access to the *data hyperuranium*, human being must take a leap of faith. Based on Promethean gap and Promethean shame, such a leap of faith takes the form of a *Promethean soteriology.* By this formula I mean the invocation of a *techno-dicy*: the only possible doctrine of salvation, once technology emerges as the only possible (last) saviour. If the Promethean *gap* generates Promethean *shame*, the latter grows into a Promethean *envy* (human being's envy of machines and artifacts, that is, of their efficiency) and then Promethean *guilt*: a negative attitude—a bad conscience—with respect to one's own "(still) being (merely) human." This guilt gives birth to the soteriological anxiety for

---

[30] Anders considers "supraliminal" those "events and actions which are too great to be understood by the human being" (Anders, 1979, 47).

[31] On this topic see (Floridi, 2014, 167–75).

a Promethean *redemption* from the only mortal sin still present in the Eden of total mobilization and total administration: *obsolescence*.

So, this leap of faith finally results in the adhesion to the dogma of "*obsolescence*" (*Antiquiertheit*), a keyword which Anders considered the anthropological benchmark of our age. According to him, obsolescence corresponds to "*man's negative attitude towards his being human*", his *voluptas* (drive) for becoming, at last, "*sicut machinae* (like machines)" (Anders, 2013, 292). In other words, to fully realize its inherent Promethean vocation, human being must reject its own humanity, must feel ashamed of itself and wish to transcend it in the form of a technological entity, that is, by turning itself into an artifact, or better, an algorithm as a coordinated set of data. By analogy to "cyborg" etymology (cybernetic + organism), in this case we could speak of "*dat-org*" (data + organism), or "inf-org" (information + organism), to use Floridi's lexicon again.[32] "We are data" in the form of "Digital Selves" (Cheney-Lippold, 2017). Seen from this hermeneutical perspective, dataism confirms to be a transubstantiation process, a reontologization of reality (Fig. 9.2).

Promethean *redemption* corresponds therefore to the commandment of "no longer being human". The self-disenchantment of human being as its adhesion to the dogma of obsolescence equates to the self-obligation to make oneself *homo materia*, so that technology may *de*-humanize us, that is, free us from the intolerable condemnation of "(still) being (merely) human". Expressed in this Promethean

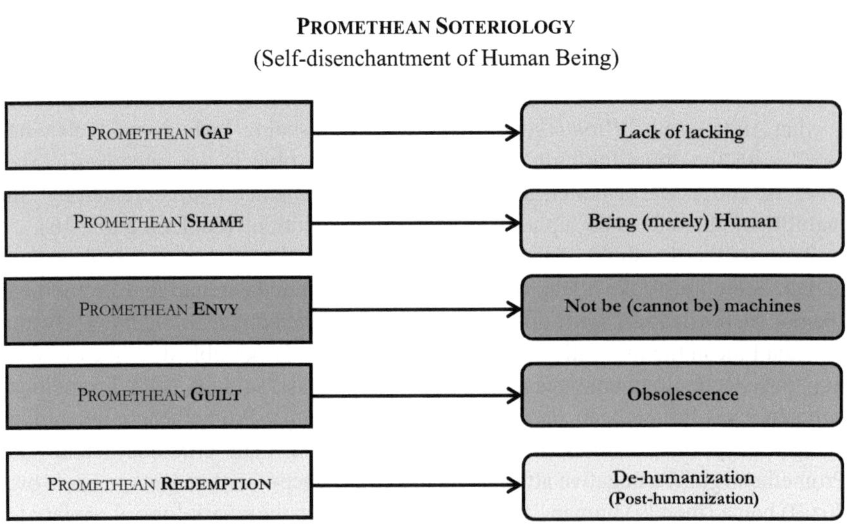

Fig. 9.2 Promethean soteriology (Self-disenchantment of Human Being)

[32] "we are informational organisms (*inforgs*), mutually connected and embedded in an informational environment (the infosphere), which we share with other informational agents, both natural and artificial, that also process information logically and autonomously" (Floridi, 2014, 94).

soteriology—in the invocation of *technodicy*[33]— is the aspiration to achieve a post-human (i.e., de-humanized) condition, which emerges as the other side (the natural *pendant*) of obsolescence. Rather than reforming the world to meet genuine human needs, it has been chosen to permanently modify human being so that it can measure up to a measure-less ("supraliminal") world. This means that, beyond appearances, the true face of a totally anthropized (i.e. *man-made*) world—a framework in which "man is the measure of all things" — is that of an "*overmanned* world."[34] Here lies the Protagoras's paradox.

Given these assumptions, the human type selected by the *technological neoenvironment*—i.e., he/she who will be allowed to cross the threshold of technological Eden—will not be a "simple" *Über-mensch*, but a real *Super-man*, that is, a literally *post*-human entity. That is to say: "*a-no-longer-(only)-human-being*". Here and now "a *better* human" means "*better than* human", namely "*more than* human". In turn, "more than human" means "*trans-/post-/meta*-human", that is, "*other than* human", "*no more* human". In its essence, as a consequence, "*trans*-humanism" means "*de*-humanization".[35] The obsolescence stands out as the symbol of a self-disenchantment of human being which proves to be self-annihilation and finally *cupio dissolvi* (wish to be dissolved). At the very moment of its definitive accomplishment, pan-anthropism proves to be a false movement: neither a dream nor a utopia, but a chimera.

On this basis, dataism confirms of not corresponding to the religion of human being (i.e., the advent of *homo deus*), rather to the religion of technology, a "techno-religion": the establishment of technology as *idolum* and object of faith. As a consequence, the age of big data—i.e., the age where the "data form" stands out as the ontological standard for all entities, the only legitimate way of being—cannot be considered an *Anthropo*-cene, but rather a *Techno*-cene. It is not the human age (Age of Humans, Human Era, *Menschenzeit*...), it is the age of technology: the epochal framework within which technology rises to the status of "subject of history" and nature as well and thus turns into an *epochal integral phenomenon*.[36]

So, the conclusion of our *critical* investigation of dataism, which aimed to enhance and rehabilitate the dimension of philosophical judgment alongside that of observation, can be summarized as follows. *Homo deus* is a mask behind which hides *homo datum*, that is, *homo materia* or human resource. In turn, *homo datum* defines himself as a *technolater*: a believer in *techne deus*.

[33] The neologism "technodicy" (*techne* + *dike*)—a further proof of the advent of techno-religion—represents the entirely secularized translation of a traditional term in theology and philosophy: "theodicy" (*theos* + *dike*).

[34] On the "overmanned world" see (Anders, 2013, 26–31).

[35] On this topic see (Müller, 2016, 99–109).

[36] On the "Technocene" see (Cera, 2023).

References

Achterhuis, H. (Ed.). (2001). *American Philosophy of Technology: The Empirical Turn* (R. P. Crease, trans.). Indiana University Press.

Anders, G. (1979). "Wenn ich verzweifelt bin, was geht's mich an?" Gespräch mit Günther Anders" [Interview with Mathias Greffrath]. In M. Greffrath (hrsg. von), *Die Zerstörung einer Zukunft. Gespräche mit emigrierten Sozialwissenschaftlern* (pp. 19–57). Rowohlt.

Anders, G. (1985). Brecht konnte mich nicht riechen. Ein ZEIT-Gespräch von Fritz J. Raddatz mit Günther Anders. In *Die Zeit*, 13/1985, 22.3.1985, 65–7—online version. Retrieved November 8, 2024, from http://www.zeit.de/1985/13/brecht-konnte-mich-nicht-riechen

Anders, G. (2002). *Die Antiquiertheit des Menschen 1. Über die Seele im Zeitalter der zweiten industriellen Revolution*. Beck.

Anders, G. (2013). *Die Antiquiertheit des Menschen 2. Über die Zerstörung des Lebens im Zeitalter der dritten industriellen Revolution*. Beck.

Asafou-Adjaye, J., Blomqvist, L., et al. (2015, April). *An Ecomodernist Manifesto*. Retrieved November 8, 2024, from http://www.ecomodernism.org/manifesto-english

Brooks, D. (2013, February 4). The philosophy of data. *The New York Times*. Retrieved November 8, 2024, from https://www.nytimes.com/2013/02/05/opinion/brooks-the-philosophy-of-data.html

Cambridge Dictionary. (2020). Retrieved November 8, 2024, from https://dictionary.cambridge.org/dictionary/english/disruptive-technology

Campagna, F. (2018). *Technic and magic: The reconstruction of reality*. Bloomsbury.

Capurro, R. (2010). Einführung in die digitale Ontologie. In G. Banse, & A. Grunwald (hrsg. von). *Technik und Kultur* (pp. 217–228). KIT.

Castells, M. (Ed.). (2004). *The network society: A cross-cultural perspective*. Edward Elgar.

Cera, A. (2017). The Technocene or technology as (neo)environment. *Techné: Research in Philosophy and Technology, 21*(2/3), 243–281. https://doi.org/10.5840/techne201710472

Cera, A. (2020). Beyond the empirical turn: Elements for an ontology of engineering. *Információs Társadalom, XX*(4), 74–89. https://doi.org/10.22503/inftars.XX.2020.4.6

Cera, A. (2023). *A philosophical journey into the Anthropocene: Discovering terra incognita*. Lexington Books.

Cheney-Lippold, J. (2017). *We are data: Algorithms and the making of our digital selves*. New York University Press.

Christensen, C. M. (1997). *The innovator's dilemma: When new technologies cause great firms to fail*. Harvard Business School Press.

Davis, E. (1998). *Techgnosis: Myth, magic, mysticism in the age of information*. Harmony Books.

Ellul, J. (1964). *The technological society* (J. Wilkinson, Trans.). Vintage Books.

Ellul, J. (1980). The power of technique and the ethics of non-power. In K. Woodward (Ed.), *The myths of information: Technology and postindustrial culture* (pp. 242–247). Coda Press.

Floridi, L. (2001). Ethics in the Infosphere. *The Philosophers' Magazine, 6*, 18–19.

Floridi, L. (2014). *The fourth revolution: How the Infosphere is reshaping human reality*. Oxford University Press.

Floridi, L. (2023). *The ethics of artificial intelligence: Principles, challenges, and opportunities*. Oxford University Press.

Frodeman, R., Klein, J. T., & Mitcham, C. (Eds.). (2010). *The Oxford handbook of interdisciplinarity*. Oxford University Press.

Galilei, G. (1960). The assayer. In *The controversy on the comets of 1618* (S. Drake & C. D. O'Malley, Trans.). University of Pennsylvania Press.

Halpern, O. (2014). *Beautiful data: A history of vision and reason since 1945*. Duke University Press.

Halpern, O., & Mitchell, R. (2023). *The smartness mandate*. The MIT Press.

Harari, Y. N. (2015). *Homo Deus: A brief history of tomorrow* (Y. N. Harari, Trans.). Harvill Secker.

Heidegger, M. (1977). The question concerning technology. In *The question concerning technology and other essays* (W. Lovitt, Trans., pp. 3–35). Garland.
Heisenberg, W. (1958). The idea of nature in contemporary physics. In *The physicist's conception of nature* (A. J. Pomerans, Trans., pp. 7–31). Hutchinson.
Hoinacki, L., & Mitcham, C. (Eds.). (2002). *The challenges of Ivan Illich: A collective reflection*. State University of New York Press.
Jaoui, A. (2000). *Le goût des autres*.
Jerónimo, M. H., Garcia, J. L., & Mitcham, C. (Eds.). (2013). *Jacques Ellul and the technological society in the 21st century*. Springer.
Jünger, E. (1998). Total mobilization. In R. Wolin (Ed.), *The Heidegger controversy: A critical reader* (pp. 118–139). The MIT Press.
Kobelieva, D. L., & Nikolaienko, N. M. (2021). From information search to the loss of personality: The phenomenon of dataism. *Anthropological Measurements of Philosophical Research, 20*, 100–112. https://doi.org/10.15802/ampr.v0i20.249591
Kranzberg, M. (1985). The information age: Evolution or revolution? In B. R. Guile (Ed.), *Information technologies and social transformation* (pp. 35–53). National Academy Press.
Lohr, S. (2015). *Data-ism: The revolution transforming decision making, consumer behavior, and almost everything else*. HarperCollins.
Marcuse, M. (2002). *One-dimensional man: Studies in the ideology of advanced industrial society*. Routledge.
Mayer-Schoenberger, V., & Cukier, K. (2013). *Big data. A revolution that will transform how we live, work, and think*. John Murray Publishers.
Mayz Vallenilla, E. (2004). *The foundations of meta-technics* (C. Mitcham, Trans.). University Press of America.
Mitcham, C. (Ed.) (1993). *Philosophy of technology in Spanish speaking countries* (Philosophy and technology Vol. 10). Springer.
Mitcham, C. (1994) *Thinking through technology: The path between engineering and philosophy*. University of Chicago Press.
Mitcham, C. (Ed.) (2005). *Encyclopedia of science, technology and ethics* (4 volumes). Thomson Gale.
Mitcham, C. (2018). Varieties of technological experience. *Issues in Science and Technology, 34*(4), 89–92.
Mitcham, C. (2020). *Steps toward a philosophy of engineering: Historico-philosophical and critical essays*. Rowman & Littlefield.
Mitcham, C. (2025). Yuk Hui's "Post-Europe": A Review. *Social Epistemology Review and Reply Collective 14*(9): 23–26. https://wp.me/p1Bfg0-agR
Mitcham, C., & Mackey, R. (Eds.). (1972). *Philosophy and technology: Readings in the philosophical problems of technology*. Free Press.
Mitcham, C., & Mackey, R. (Eds.). (1973). *Bibliography of the philosophy of technology*. Chicago University Press.
Morozov, E. (2013). *To save everything, click here: The folly of technological solutionism*. PublicAffairs.
Müller, C. J. (2016). *Prometheanism technology: Digital culture and human obsolescence*. Rowman & Littlefield.
Nietzsche, F. (2006). *Thus spoke Zarathustra: A book for all and none* (A. Del Caro, Trans.). Cambridge University Press.
Nietzsche, F. (2017). *The will to power: Selections from the notebooks of the 188os* (R. K. Hill & M. A. Scarpitti, Trans.). Penguin Books.
Noble, D. F. (1999). *The religion of technology: The divinity of man and the spirit of invention*. Penguin.
Plato. (1921). Sophist. In *Plato II: 'Theaetetus', 'Sophist'* (The Loeb Classic Library) (H. N. Fowler, Trans., pp. 259–459). Heinemann and G.P. Putnam's Sons.

Romele, A. (2023). The datafication of the worldview. *AI & Society, 38*, 2197–2206. https://doi.org/10.1007/s00146-020-00989-x

Snow, C. P. (2012). *The two cultures*. Cambridge University Press.

van Dijck, J. (2014). Datafication, dataism and dataveillance: Big data between scientific paradigm and ideology. *Surveillance & Society, 12*(2), 197–208.

Vico, G. (1971). De antiquissima italorum sapientia ex linguae latinae originibus eruenda. *Opere* (P. Cristoforini, Italian Trans., pp. 55–131). Sansoni.

Weber, M. (2012a). The meaning of 'value freedom' in the sociological and economic sciences. In H. H. Bruun and S. Whimster (Eds.), *Max Weber: Collected methodological writings* (H. H. Bruun, Trans., pp. 304–334). Routledge.

Weber, M. (2012b). Science as a profession and vocation. In *Max Weber: Collected methodological writings* (pp. 335–353).

Wheeler, J. A. (2018). Information, physics, quantum: The search for links. In W. H. Zurek (Ed.), *Complexity, Entropy and the Physics of Information: The Proceedings of the 1988 Workshop on Complexity, Entropy, and the Physics of Information Held May–June, 1989, in Santa Fe, New Mexico* (Vol. 8, pp. 309–336). CRC Press.

Part V
Retrospective on Mitcham's Contribution to Dialogue with China

Chapter 10
Philosophy of Technology and Engagement with China: Interview with Carl Mitcham

Zhihui Zhang

Abstract Since the early 1990s, work by Carl Mitcham has been influential in the development of Chinese Science, Technology, and Society (STS) studies and philosophy of technology. This interview examines the intellectual trajectory of Carl Mitcham's philosophical development and his connections with China. It provides some biographical background, reviews the history of the Society for Philosophy and Technology (SPT) and the Forum on Philosophy, Engineering, and Technology (fPET), and identifies some key points in his engagements with Chinese scholars such as Yin Dengxian (Chinese Academy of Social Sciences), Chen Fan (Northeastern University), Li Bocong (University of Chinese Academy of Sciences), Wang Qian (Dalian University of Technology), and Liu Yongmou (Renmin University of China). The interview includes an examination of some specific developments of philosophy of technology in the West: the ethics of technology and the policy and empirical turns in the philosophy of technology as well as Mitcham's personal description of what he has learned from Chinese philosophers.

Keywords Applied ethics · Chinese philosophy · Empirical turn · Ethics of technology · Philosophy of engineering · Philosophy of technology · Policy turn · Science technology and society (STS) studies

10.1 Biographical Background

Zhihui Zhang: As one outcome from my time as a visiting scholar with you at the Colorado School of Mines, I would like to conduct an interview to offer some insight into your work, especially for Chinese scholars and students. As you know,

---

This interview was made by Zhihui Zhang with Professor Carl Mitcham in 2014, when she was an visiting scholar at Colorado School of Mines for several months.

Z. Zhang (✉)
Institute for the History of Natural Sciences, Chinese Academy of Sciences, Beijing, China
e-mail: zhangzhihui1008@163.com

L. Checketts (ed.), *What Went Wrong with the West: Carl Mitcham's Critique of Western Philosophy of Technology*, Philosophy of Engineering and Technology 52, https://doi.org/10.1007/978-3-032-22214-5_10

after your first visited China in 1992, you became well known as both a Science, Technology, and Society (STS) scholar and philosopher of technology.

Carl Mitcham: It has been a pleasure getting to know you and your work, especially on the Three Gorges Dam. I'm honored by your interview proposal.

ZZ: Let's begin with your background. I know you received your bachelor's and master's degrees in philosophy from the University of Colorado, then eventually a doctorate in philosophy from Fordham University in New York. But I wonder what originally made you interested in philosophy, especially the philosophy of technology.

CM: Philosophy has been important to me from almost as long as I can remember. Part of what it means to be human is to ask philosophical questions, questions about who we are and how we should act, and all of us do this to some extent. For some people, however, it becomes central to how they lead their lives. Even in primary school, I wondered about language and how words came to be. One time I asked my Mother, "Why is a washing machine called a 'washing machine' and a stove a 'stove'?" Another time I asked my father about what made noise into music. I have a tin ear and didn't think there was anything all that great about classical music, which my father liked. Violin music seemed not really all that different from screeching noise.

It was probably at age 12 or 13 that I first started reading books with "philosophy" in the title. I remember purchasing Bertrand Russell's *A History of Western Philosophy* (1945) and reading it, not understanding much, but getting interested and then going on to read some of his other books. Russell's history of philosophy is not very profound, but it had the beneficial effect of introducing me to many names and ideas.

ZZ: And what about the philosophy of technology? I don't think you can find much of this in Bertrand Russell.

CM: True, Russell never uses the term "philosophy of technology," but he is not as oblivious to technology as many other analytic philosophers. He became a philosophical critic of nuclear weapons and participated in the "ban the bomb" movement in the UK in the late 1950s and early 1960s. At age 89 he was even arrested and spent 7 days in jail for participating in a large anti-nuclear protest in London.

So Russell had an influence on me as a model for how philosophy is more than an academic profession, as something that has implications for how we lead our lives. Philosophy is not separated from the world; it is a particular kind of engagement with the world.

ZZ: Did you read any American philosophers too?

CM: Yes. Another early influence was the radical American psychologist and philosopher Paul Goodman. (Most academic philosophers would not consider him a philosopher.) Goodman's *Growing Up Absurd* (1960), which I read shortly after it was published, pointed toward technological change as undermining relationships between people. Because of technological change, for instance, a young boy could no longer expect to do the same kind of work as his father; he could no longer take his father as a role model. Tradition, which had been the way people linked themselves to the past and to a larger reality, was being replaced by "technological progress"—a concept Goodman questioned. Goodman even suggested that

European existentialism and the concept of the absurd were products of a technological distortion of the world. So from the beginning, trying to think about this world transformation was what philosophy of technology meant to me.

Incidentally, Goodman's essay "Can Technology Be Humane?", which argued that "technology is a branch of moral philosophy, not of science," is an underappreciated classic in American philosophy of technology.

ZZ: So the atomic bomb was an issue for you?

CM: Yes. I was very aware of the atomic bomb. Goodman writes about it in the article I just mentioned—more insightfully than Russell, I might add.

I remember one night when I was 6 or 7 years old, waking up in the middle of a thunderstorm with lots of lightning, and thinking that the Soviet Union was bombing us with nuclear weapons and that I was going to die. I called out to my mother asking if these were atomic bombs. In my primary school we practiced "duck and cover" drills in case of an atomic war, and the idea of unprecedented destruction was a persistent presence in my mind. Such ideas were part of American popular culture in the form of grade-B movies like *Them!* (1954) and eventually the post Cuban Missile Crisis classic *Dr. Strangelove* (1964).

ZZ: Did your parents talk about these things? What was their attitude toward Russia—and toward Communism and China?

CM: I was born the year Japan bombed Pearl Harbor and the United States entered World War II. My parents told me that I was lying on a rug in our living room when news came over the radio that we had been bombed. My father, who was a mechanical engineer, tried to join the army but was asked instead to teach engineering to young soldiers at a local military college.

During the war, the Soviet Union had been an American ally, but shortly after war ended in 1945, the Cold War began. My parents, like most people in the United States, became strongly anti-Soviet and anti-Communist—and after 1949, anti-Chinese as well. A novel (subsequently a movie) that crystalized the American fear of Communism was Richard Condon's *The Manchurian Candidate* (1959). In it an American soldier captured during the Korean War was taken to Manchuria, where Chinese brainwashing techniques turned him into a secret Communist agent. After he was released and returned to the United States, he was supposed to become a Presidential assassin. The book was made into a movie not once but twice (1962 and 2004), both times expressing and cultivating fear of Communism and China.

I should add that my parents thought education was very important and encouraged me to read and think, even though my thinking eventually led to deep disagreements with them. When I was in primary school they bought a set of books called the *World Book Encyclopedia* so I could look up and study anything I wanted. This was before the advent of the Internet, when reading books was still essential. My father had a set of *Harvard Classics*, a 51-volume anthology of great books from mostly Western history. The only writing from China was "Sayings of Confucius" in volume 44. At one time I had a goal of reading the whole set but never actually got very far.

ZZ: What kind of work did your parents do?

CM: My mother was initially a stay-at-home mom, but after my younger sister started school, she began teaching. After the war, my father got a job as a mechanical engineer with Otis Elevator Company. Both parents were also Protestant Christians. We went to church every Sunday. Religion was a serious influence, although as a teenager I became skeptical of the Christian claims to revelation.

ZZ: What about your university education?

CM: I'd prefer to skip over this for now. Suffice it to say that for me it took longer than usual. As you already noted, I earned a double BA in Philosophy and Liberal Studies from the University of Colorado, continuing there for an MA. Then some years later I earned a PhD in Philosophy from Fordham University, a Jesuit school, in New York City. This took place over a 30-year time span. By the time I was awarded a doctorate I had already edited (with my long-time friend, Robert Mackey) *Philosophy and Technology* (1972) and become a tenured professor at Brooklyn Polytechnic University, thanks to the good offices and patronage of George Bugliarello, a civil engineer and its president, who was an engineer philosopher. My academic career has been unusual.

During this whole period I was completely oblivious to what was happening in China—and pretty much even the very existence of China. I was not interested in Chinese history, culture, literature, or philosophy. The fear of China I learned as a child may have dissipated, but it was not replaced by any positive interest.

10.2 From the Society for Philosophy and Technology (SPT) to the Forum on Philosophy, Engineering, and Technology (fPET)

ZZ: OK, Let us move forward to 1978, the period of the Reform and Opening and the year that the Society for Philosophy and Technology (SPT) was created and you were elected its first president. When SPT began editing its journal, *Research in Philosophy and Technology,* you became the associate editor. This was an important step in the institutionalization of philosophy of technology. How did you become involved?

CM: Paul Durbin was the creative impetus for creating the Society for Philosophy and Technology. He was my mentor—although I never fully appreciated him until after his death in February 2019, when I organized a small memorial symposium on his work for the 2019 SPT biannual meeting. An expanded version of this symposium has recently been published (Mitcham, 2020).

ZZ: How did you and Durbin meet?

CM: About 1971, when Mackey and I sent our typescript for the *Bibliography of the Philosophy of Technology* (1973) to Melvin Kranzberg, the editor of *Technology and Culture*, he sent it out for review. Durbin happened to be one of the reviewers. He wrote a strongly positive review and broke cover to get in touch directly, saying in his typically straightforward and unpretentious manner, that we should have included a couple of his articles. We began corresponding—the old fashioned way, through written letters exchanged by post.

I was living in Kentucky at the time, teaching at a small, private liberal arts college (Berea College, 1970–1972) and then at the even smaller, Catholic institution (St. Catharine College, 1972–1982). Durbin, a full professor at a major public research institution, invited me to visit the University of Delaware. I didn't yet have a PhD, but that didn't bother him. Right from the start he was friendly and supportive. Later when Jim Grote (another Kentucky friend) and I made a research visit to the Library of Congress in Washington, DC, Durbin invited us to stay at his house in nearby Wilmington, Delaware, sleeping on his living room sofa, meeting his wife and family.

In the course of our correspondence, Durbin came up with the idea for a conference. I had only been to one previous philosophy conference and had no idea how to organize such a thing. But he invited me to help with the first American philosophy conference on technology at the University of Delaware in 1975. I was able to identify a few people to invite. One of the most important was Albert Borgmann, whose work I had recently discovered in the course of bibliographic research.

As an aside, I should add that 2 years earlier (in 1973) Bugliarello, then Dean of Engineering at the University of Illinois Chicago Circle, had organized an international conference on the history and philosophy of technology to which Mackey and I had been invited, but which mistakenly failed to include Durbin. Bugliarello invited mostly engineers and historians along with a few philosophers. Durbin's conference was only philosophers. But the edited volume that came out of Bugliarello's conference (Bugliarello & Doner, 1979) should not be overlooked in any account of an American emergence of the philosophy of technology.

As Durbin wrote in 1978, out of the 1975 "Delaware conference and a follow-up symposium at the American Association for the Advancement of Science meeting in Boston, February 1976" came *Research in Philosophy and Technology* (vol. 1, 1978), plans for future conferences, and a newsletter. Durbin, as general editor, invited me to work with him as review and bibliography editor.

ZZ: So when was SPT formed? The title page of *RPT* vol. 1 describes it as the "official publication of the Society for Philosophy and Technology."

CM: Probably in 1977, at least informally. In order to make it more formal, Durbin used the occasional *SPT Newsletter* to create an *ad hoc* nominating process that put me on a ballot with Mario Bunge. Then a similar *ad hoc* voting process elected me with perhaps one or two more votes (out of perhaps 20 total) than Bunge. A first task was to draft some by-laws. One evening in 1979 or 1980, after a panel at a Philosophy of Science Association meeting (if I recall correctly) that brought the newly elected SPT Secretary-Treasurer Stanley Carpenter from Georgia Tech to Louisville, Kentucky (near where Jim Grote and I were living), a few of us got together in the living room of Professor Kristin Sharader-Frechette (who was teaching philosophy at the University of Louisville). Using as models the by-laws of the Society for the History of Technology and the Philosophy of Science Association, we drafted the SPT by-laws.

At the election it was not clear what my term of service would be. The new by-laws stipulated it would be for 2 years, but the beginning date got played forward to 1981–1983, so as to coincide with the first official SPT international conference, hosted by Friedrich Rapp in Bad Homberg, Germany. (Because I missed an airline flight, I failed to make the Bad Homberg gathering.)

At that same time Bugliarello, whom I mentioned earlier, after moving from Chicago to become President of Brooklyn Polytechnic University, brought me to this technological institution to help him establish a new Philosophy and Technology Studies Center. This provided a platform for organizing the second official SPT international conference there in 1983.

ZZ: What about the Forum on Philosophy, Engineering, and Technology (fPET)? Weren't you also involved in the setting up of this organization? How does it differ from SPT?

CM: I think of myself as equally involved in fPET and SPT, but the existence of fPET must be credited to Taft Broome, an African American Professor of Mechanical Engineering at Howard University in Washington, DC. Broome had done some work in engineering ethics, but then at a certain point in his career took a sabbatical and went back to university to earn a master's degree in STS from Rensselaer Polytechnic Institute. This was an unusual and rather brave move.

In the process Broome became convinced of the need for philosophy of engineering as something distinct from philosophy of technology, as a way to help engineers understand themselves. He thought engineers and engineering were too marginalized in philosophy of technology circles. He was right about this. The fact that engineer Bugliarello had organized a philosophy of technology conference in 1973 and philosopher Durbin another in 1975 with little overlap in contributors reflected a separation between these two communities and interests.

Broome eventually received modest encouragement from the U.S. National Academy of Engineering to create a small working group to pursue his idea. He then organized an invitational Workshop on Philosophy and Engineering (WPE) at MIT in 2006. I was invited, along with 10–15 others. We sat around a table for a day and agreed to endorse Broome's proposal for a larger workshop, which Delft University of Technology (TU Delft) offered to host the following year. That same year there was a special session on philosophy of engineering at SPT 2007 in Charlestown, South Carolina, in which a few of us proposed merging WPE with SPT. However, Broome felt strongly the need for some engineering independence, and the following year a second WPE (2008) was hosted by the Royal Academy of Engineering in London.

At WPE 2008 some disagreements emerged about what should come next, the outcome of which was a modest re-organization as fPET and a decision to hold conferences every other year, on even numbered years, in complement to SPT conferences on odd numbered years. I then hosted the first fPET conference at the Colorado School of Mines in 2010.

As you know, Professor Li Bocong and the Graduate University of Chinese Academy of Sciences hosted fPET 2012 in Beijing. After a 6-year delay, a proceedings volume of that conference was published under the co-editorship of Li Bocong, Byron Newberry, Zhang Baichun, and myself (2018).

Perhaps this is sufficient institutional background for now, don't you think?

10.3 Engagements with China

ZZ: I would like to ask now about your relationship with China. As I noted at the beginning of our interview, you are most well known in China as an STS scholar and philosopher of technology. How did it happen you became involved with Chinese scholars working in these areas?

CM: Like many things in my career, it was not planned. (I never actually planned to become a professor.) As already mentioned, for many years I had no interest in China; the connection came quite late in my career, almost like a second life. It emerged because China first expressed interest in me.

My first contact with a Chinese philosopher of technology came by way of a letter dated January 12, 1987, from Professor Gao Dasheng, who wrote to me at Brooklyn Polytechnic University, where I had become Director of the new Philosophy and Technology Studies Center. I'm not sure exactly how he identified me, but it may have been the result of a special issue of the journal *Technology in Society* (1984) which I had guest edited on "Philosophy and Technology." Bugliarello was founder of the journal and saw this special issue as a way to promote the new Studies Center. Perhaps it would be appropriate to quote the first two paragraphs of the letter from Gao:

> Dear Prof. Mitcham:
>
> I'm the director of History and Philosophy of Science Division, Social Science Department, Tsinghua University. In 1985, as a representative of China, I attended the 17th International Conference of History of Science held in Berkeley. The beautiful scenes and enthusiasm people of America had impressed me deeply.
>
> Tsinghua University has been a leading university in China since its birth, especially in technology and engineering. In such a university, my colleagues and I intend to build an interdisciplinary unit to study technology from the view of philosophy, history and sociology. I hope to establish academic relationship with you, and to get directions and help from you.

The letter concluded by asking if I might be able to help acquire some publications in the philosophy of technology that were not available in China. I responded by sending publications and inviting him to write an article on philosophy of technology in China for a future issue of *Research in Philosophy and Technology*. Gao responded in October 1987 with an article co-authored with his student Zou Tsing, which I edited for publication in *RPT*, vol. 6 (1989). This article provided a first introduction for English readers to the thought of such Chinese philosophers as Chen Changshu, Liu Zeyuan, and Yuan Deyu. At this point, however, I did not fully recognize what had been introduced to me; my contact with China remained minimal. I regret that I never met Professor Gao or his student.

The next year Zou Tsing was awarded a Young Chinese Program Fellowship to study with me at Pennsylvania State University (where I had moved), but in a letter dated September 21, 1990, he wrote that "with great regret" he was not able to take advantage "owing to unexpected circumstances and the change of my plan."

Nine months after that first letter from Gao Dasheng, I received another surprise note from Meng Qingshi with a copy of his translation of my 1985 article "What Is the Philosophy of Technology?" in哲学译丛*Zhexue Yigong*. The note, dated September 29, 1987, simply said "Presented with pleasure to Professor Mitcham," adding that the publication venue was "a Chinese journal dedicated only to introduce foreign philosophies among the world, and edited by the Institute of Philosophy Research, the Chinese Academy of Social Sciences." I regret never having had occasion to meet Meng Qingshi.

I first visited in 1992, when another scholar from the Chinese Academy of Social Sciences, Professor Yin Dengxiang, prevailed on Professor Stephen Cutcliffe and me to come and give a series of lectures introducing STS. At the time Cutcliffe was Director of the Lehigh University Science, STS Program and I directed the Pennsylvania State University STS Program. (The two universities are about a hundred miles apart.) Yin had spent some time as a postdoctoral scholar with each of us the year before.

Despite previous contacts with Gao Dasheng and Yin Dengxiang I was not yet seriously interested in China; I was much more attracted to developments in the philosophy of technology and STS that were taking place in the Netherlands and in Spain. Spain and the Spanish-speaking world was the non-Anglo culture to which I was most attracted as a counterpoint to the shallowness of American culture and history. China was just too foreign.

Nevertheless, after some hesitation, Cutcliffe and I agreed to make the trip and prepared a combined 4 days of talks, each of us speaking half a day in rotation. This took place in June 1992 at the Graduate University of Chinese Academy of Sciences in Beijing. I had never flown so far before and it was a hard trip. Not to mention my total inability to speak or read Chinese, it was difficult to adjust to Chinese food and culture. At that time Beijing was much more polluted than today and the streets were full of bicycles.

Your question here has stimulated me look to back through my files, where I have discovered notes and photographs that make me nostalgic. Perhaps on another occasion we could look through these together. (Together with a collection of books in the philosophy of technology, I am depositing some of these records in the library at Renmin University of China.)

ZZ: What did your and Professor Cutcliffe's lectures cover?

CM: Day one was a general introduction followed by a review of the history of technology in the West. Day two focused on relationships between technology and society, along with a case study in environmental history. Day three was devoted to politics and technology. Day four dealt with ethics and technology, and with STS pedagogical practices. Looking back, I'm surprised there was no lecture on explicitly centered on the philosophy of technology, although for me philosophy of technology (or "philosophy and technology studies," my preferred term) includes all the topics mentioned.

After our talks in Beijing we were flown south to give a few talks at Nanjing University, but then returned to Beijing by rail in order to get more of a feel for the countryside. This was before bullet trains, and the journey took more than 24 h. The experience has been briefly described by Cutcliffe (1993). Although I had originally resisted China, I found myself impressed by the friendliness of Chinese people, the dedicated seriousness of Chinese scholars, and their desires to learn the state of scholarship in the United States. The seeds for my Sinophila had been planted.

ZZ: So after that you started visiting China on a regular basis?

CM: Not yet. My second visit did not take place until 6 years later.

During the 1992 visit, Cutcliffe and I met a number of scholars whose importance I did not appreciate until much later: Among these was Chen Changshu, Chen Fan, and Li Bocong. I am especially embarrassed now not to have talked more deeply with Professor Chen Changshu. Subsequently, Professor Li Bocong made a visiting scholar trip to the United States during the 1994–1995 academic year, again visiting both Cutcliffe at Lehigh University and me at Pennsylvania State University. Additionally, Professor Yin Dengxiang undertook to translate my short introduction to the philosophy of technology, which had been published in Spanish in 1989. I provided the original English text, and with work by Professor Cao Nanyan from Tsinghua University, the Chinese version appeared in 1998.

In conjunction with this publication, Yin Dengxiang invited me to make a return visit. Two years before (in 1996) he had invited me to participate in a Sino-American Symposium on Humanistic Factors in the Development of Advanced Science and Technology hosted by the Chinse Academy of Social Science, but I had been unable to attend. Now, remembering how hard the original trip was, I resisted again. The seed of Sinophilia had not yet fully sprouted. But Yin watered it with encouragement—after all, this would be an occasion for promoting the Chinese translation of my introduction to philosophy of technology—and I visited for a second time in June 1998 for a Sino-American Academic Symposium on Humanistic Factors in High Science-Technology. There were approximately 40 participants, but the only foreigners were Hans Poser from Germany, Albert Moyer from Virginia Tech, Steve Cutcliffe, and me.

ZZ: How did what you call your Sinophilia finally emerge?

CM: Now two Chinese scholars had come to visit and I had twice visited China. Other scholars and students began to contact me. Wang Nan, now an Associate Professor at University of Chinese Academy of Sciences, emailed me in 2004, when she was still a graduate student at Dalian University of Technology. She asked for help in locating some materials on the work of Ernst Kapp and became one of my longer term Chinese collaborators. Then in 2007 Professor Wang Qian invited me to participate in a conference at Dalian University of Technology where I met Zhu Qin, who became my first predoctoral visiting scholar. You and I also met about this time. Since then I have traveled to China almost every year and visited a number of universities in cities from Harbin and Shenyang to Kunming and Guangzhou, from Dalian and Shanghai to Changsha.

The year 2011 was another pivotal year because for the first time I began to make a serious effort to appreciate Chinese history and culture. At the invitation of Professor Cao Nanyan (and with Liao Miao as my TA), I taught a 5-day seminar on Science, Technology, and Ethics at Tsinghua University. Afterward Wang Nan offered to serve as a guide and took me to visit the Confucian temple at Qufu and then climbed Taishan. It is difficult for me to describe the personal significance of that experience. It was where the seed of Sinophilia began to blossom. It was also at Tsinghua that July that I first began to try to learn a little Chinese, when Professor Cao Nanyan arranged for a graduate student from Beijing University to serve as a tutor. The tutor, whose name I wish I remembered, gave me a copy of Lu Xun's *Call to Arms* (呐喊), thereby introducing me to modern Chinese literature.

Thus when, in 2014, Professor Liu Yongmou approached me about serving as a regular Visiting Professor in the School of Philosophy at Renmin University of China I readily responded with a desire to do so. The faculty and students at Renmin University have been more welcoming than I could possibly imagine. Renmin University has become a second home. But I continue as well to be indebted to many other scholars, institutions, and students in China.

ZZ: Can you say something concretely about what attracts you to China?

CM: Unfortunately, I'm afraid at first there was a strong element of what Edward Said (1978) terms "Orientalism": I imagined Chinese scholars as more receptive than the United States to my criticisms of engineering and technology. I thought China would be able to avoid some of the mistakes America was making with technology, that China might pioneer an alternative path in technological development. As more than one Chinese scholar has gentle chided me, I tended to romanticize China. But I have also discovered among Chinese a counter "Occidentalism" that projects onto the United States some romantic illusions of their own. In the language of chaos theory, Americanism functions as a strange attractor in a complex global political system. (I suspect that the presidency of Donald Trump is correcting Chinese romanticism about America.)

Instead, paradoxically, one thing that I reluctantly learned from China is an appreciation of the inherent attractiveness of technology—and especially why it exercised such an attractiveness for Americans during the raging, chaotic hundred years from 1850 to 1950. Its bright promise became, as the philosopher Li Zehou would say, sedimented in American culture, or, in the words of Albert Borgmann, infused its character into contemporary life—to such an extent that, as my mentor Ivan Illich (1972) has argued, it often became counterproductive.

In China I have found myself stepping back into the heroic age of technology and witnessing some of its commanding features—features nonetheless slightly more separated from warfare, violence, and destruction than was the case during the American heroic age. In China, it does appear that technology has served a less aggressive role of freeing the country from Western imperialism and liberating hundreds of millions from poverty—although its bright promises are not wholly free from dark sides.

ZZ: I have heard you argue that just as Chinese have been learning from the West, the West, and especially Americans, need to learn from China. In light of this new perspective, do you still think Americans can learn something from China? When did you start thinking that Americans should learn from China?

CM: Yes, I continue to think Americans have much to learn from China. In the first place, Chinese students are generally more curious about America than American students are about China. The same applies to Chinese scholars and maybe even to the Chinese public as a whole. At the 2014 annual meeting of the

American Society for Engineering Education I gave an invited talk, "Learning from China," that called attention to how many more Chinese students are studying English than American students studying Chinese. How there is an order of magnitude more Chinese studying in the United States than American students studying in China. How willing students in China are to listen to visiting professors lecture in English or even French and German. (American students would never go to lectures in a foreign language.) How Chinese students in general are more serious and harder working and certainly less privileged than Americans. China's determination to learn from the United States should be a model for the United States to learn from China. China has things to teach Americans and the world.

One thing that is distinctive about China is not only that it has the longest continuous civilization in world history; it has a unique history of decline and fall—and revival. China has declined and fallen many times, but unlike Rome it has also repeatedly reconstituted itself, exhibiting a kind of historical *élan vital*. Facing the possible catastrophe of climate change, this may be an experience on which more than China will be called to draw.

10.4 From Philosophy to Ethics of Technology: 1970s to 1990s

ZZ: It surprises me also that in 1992 you were not asked to give a talk on philosophy of technology. Perhaps it was because you came to China as the director of an STS program and your major philosophy of technology book, *Thinking through Technology* (1994), had not yet been published.

CM: Perhaps, but the two books I did with Robert Mackey on philosophy of technology were 20 years old by then, and I considered philosophy and technology studies my primary contribution to the STS field. At that time STS was often conceived as an interactive or interdisciplinary combination of the history and philosophy and sociology of science and technology: a total of six disciplines.

Note too that I preferred the term "philosophy and technology" to "philosophy of technology." I have disagreed with both Don Ihde and Joseph Pitt about this, in order to stress the need for philosophy to collaborate on a more equal footing, as it were, with technology. In my lectures, day three was on politics and technology; day four on ethics and technology. At that time, political philosophical and ethical issues were the most prominent philosophical issues related to technology. At the 1998 symposium, my presentation again was on technology and politics and the need for public participation in technoscientific decision making. This is a topic on which my view has continued to evolve, especially in a society where public participation has been turned into a weapon against appreciation of expertise.

ZZ: How do you see the relationship between ethics, politics, and more general philosophy of technology?

CM: My first published article with the titled "Philosophy of Technology" was an entry in the original edition of the *Encyclopedia of Bioethics* (1978). A revision prepared for the second edition (1995) was translated into Chinese and included in the booklet that accompanied my 1992 lectures. A later entry with the same title for the second edition of the *Encyclopedia of Philosophy* (2006), described the relationship this way:

> The philosophy of technology brings logical, metaphysical, epistemological, ethical, and political philosophical questions to bear on the making and using of artifacts. The particular balance among these questions will differ within related regionalizations of philosophy, such as the philosophy of science or the philosophy of art. In the philosophy of technology, for instance, epistemology typically plays a lesser role than in the philosophy of science but a greater role than in the philosophy of art. Any philosophical assessment of technology is thus partially defined by its own inner balance in relation to philosophy as a whole. (Mitcham, 2006, p. 543)

This was actually reflected in both of my 1970s books on philosophy and technology. The largest sections of the 1972 collection and 1973 bibliography were on "Ethical and Political Critiques," but both volumes also included sections and articles on conceptual analysis, epistemology, religious perspectives, and metaphysics. I thought that even when there was an emphasis on ethics driven by social concerns, philosophy and technology studies had to bring other branches of philosophy to bear.

ZZ: So for you, ethics of technology was just the most pronounced aspect of the philosophy of technology, reflecting the prominence of technology as a social issue. But even when you did work on ethics and technology, you were doing it as part of the philosophy of technology.

CM: Correct. It also helps to place ethical discussions of technology in the context of the 1970s "applied turn" in philosophy. From the early 1970s American philosophers became interested in trying to bring philosophy to bear on any number of contentious social issues. Although the movement was often called the "applied turn" in philosophy—an edited volume on *The Applied Turn in Contemporary Philosophy* (Attig et al., 1983) provides an overview—the applied turn was heavily weighted toward ethics. Although in principle applied philosophy includes applied metaphysics and epistemology, in fact public policy and ethical issues were front and center. The journal of *Philosophy and Public Affairs* (founded 1972) provides a good example: questions about just warfare, abortion, and criminal punishment were repeated themes of early issues. Even in the *Journal of Applied Philosophy* (founded 1984) ethics and political philosophy were more prominent than ontology, epistemology, or aesthetics.

There was nevertheless a failure in these developments in applied ethics to appreciate the degree to which many such issues underwent mutations created by modern technology. The question of when warfare can be justified has been amplified by the creation of nuclear weapons and other advanced technological armaments. As population increases in the wake of advances in public health and medicine, societies started asking questions about birth control. Advanced lifesaving medical technologies have stimulated questions about how to define death. In standard applied ethics, the technological dimensions of many issues were not always adequately addressed.

ZZ: I notice from your CV that in the late 1980s you edited a book on *Ethics and Technology* (1989), co-edited another on *Ethical Issues Associated with Scientific and Technological Research for the Military* (1989), and the same year as you first visited China edited still another on *Engineering Ethics throughout the World* (1992). You also published articles on computer ethics and biomedical ethics and a short, co-authored volume on *Engineering Ethics* (2000). Then during the next decade you edited the four volume *Encyclopedia of Science, Technology, and Ethics* (2005) and won the International World Technology Network (WTN) award for Ethics (2006).

CM: My interest in all these instances was to try to expose the ways in which ethics and politics today have become wrapped up with technology. In 1996 Leonard Waks and I co-authored an article arguing this general point. The article began by pointing out how Alasdair MacIntyre in *After Virtue* (1981), while correctly noting the inadequacies of contemporary philosophical reflection on nuclear weapons, abortion, and healthcare overlooked a salient feature: the extent to which their new ethical challenges could be traced to transformations brought about by advances in technology. My privately published book *Thinking Ethics in Technology* (1997) tried to make the same case. (As an aside, this volume was meant as the first draft of a book to be called *High-Tech Ethics: Learning to Live with Advancing Technology* that unfortunately never appeared. Before he became ill, I corresponded with Chen Changshu about doing a Chinese version.)

ZZ: Is there a reason why your work during this time was so devoted to editing and was so collaborative? You seem to have co-authored and co-edited more than is usual for philosophers.

CM: My editing reflects, quite frankly, a sense that in many instances other scholars are more insightful than I am and a desire to learn from them while making their work more generally available. Collaborating sometimes helps me finish projects from which I might otherwise get distracted. I have an unfortunate tendency to start many projects that I fail to finish. (Our interview here, which has taken place over a number of years, is another example.)

10.5 Ethics and Policy in Engineering and Technology

ZZ: Since you have mentioned the applied turn in philosophy, maybe you could say something about what you have called a "policy turn" in philosophy of technology. You gave a talk at fPET 2012 in Beijing on engineering policy that tried to extend the notion of science policy. You spoke about this topic again at SPT 2015 in Shenyang. But people have been unsure about what you mean.

CM: My use of the term "policy turn" grows out of collaboration with Robert Frodeman, who was a visiting professor at the Colorado School of Mines in

2001–2002. Frodeman had worked with the U.S. Geological Survey and as a faculty member in the Center for Science and Technology Policy Research at the University of Colorado in Boulder. We wound up talking a lot about policy, what it is, and how it is different from both ethics and politics.

"Policy" is a strange word—and a fuzzy phenomenon—that I had been struggling with from my time working with Ivan Illich at Pennsylvania State University in the 1990s. Illich thought it was just another example of "uniquack" (see Illich & Sanders, 1989) and called it an "amoeba word," adapting his friend Uwe Poerksen's (2004) concept of "plastic words."

I thought it was not so easily dismissed, and collaboration with Frodeman led to co-editing a theme issue of the journal *Philosophy Today* titled "Toward a Philosophy of Science Policy: Approaches and Issues" (2004) in which we sought to nurture philosophical reflection on the topic. The year before, Frodeman and I had co-authored with science policy researcher Roger Pielke Jr. (2003) a short piece on "humanities policy."

ZZ: So what is your view about the meaning of policy?

CM: For Pielke, a leading scholar and practitioner in the field of science policy, a policy is just a decision: nothing more, nothing less (Pielke, 2007). This seems to me an inadequate characterization. The fact that we can talk about "policy decisions" in contrast to economic, political, or ethical decisions suggests that policy is something distinct.

Policy provides a distinctive kind of guidance for human action that is different from statutory law, military command, moral duty, technological rule, and more. One dream of the European Enlightenment was to replace political commands based on force, power, tradition, or superstition with, insofar as possible, conduct grounded in scientific knowledge. The ideal of modern science has been to replace superstition with knowledge, including knowledge as a guide to action. For instance, instead of deciding the treatment for a medical illness on the basis of intuition, tradition, or superstition, we want to decide on the basis of empirical evidence about what works and what doesn't. Insurance policies (as opposed to simple bets) are constructed using statistical analysis of data about the occurrence of certain events, not just guesses or irrational fears.

Science policy is commonly divided into two types: *science for policy* and *policy for science*, in a distinction often credited to science policy adviser Harvey Brooks (1964). The former brings science to bear on political decision making in order to turn politics into policy, replacing oracle bone based decisions or intuitions with scientifically based assessments. Governments that want to provide safe drinking water need to utilize scientific research to provide knowledge about levels of contaminants and to engineer effective processes for cleaning up impure water. The latter seeks to establish guidelines for the support and governance of science, on the basis (again) of empirical research. How much money should a government invest in scientific research and how is it most effectively invested? Economics and other social sciences should be used to establish of a feedback loop of science for policy for science.

Now two things can be noted in this description of science policy. First, it is as much about engineering as about science. Both medicine and the designing of safe drinking water systems are forms of engineering as much if not more than science. Second, policies complement ethics in providing guidance for action. Ethics focuses on individual decision making and actions. Policies provide guidance of a more general sort for groups or people occupying distinctive roles in specific social contexts. Insurance policies do not leave it up to the guess work or the personal preferences of individual insurance agents to write contracts: Instead, they specify the parameters of a contract, on the basis of statistical analyses. Physicians have a moral obligation to do no harm. But what counts as harm needs to be based at least in part on empirical research into the likely outcomes of various treatments, especially in a world such as ours with harms that can increasingly result from technoscientific inventiveness.

In short, in engineering and technology, ethics alone is not enough. Guidelines for moral conduct in the engineering design and use of technologies need to be based on more than consequentialist, deontological, or virtue ethical moral reasoning. Moral reasoning needs to be complemented with what Harold Lasswell (1951) called the policy sciences.

Frodeman and I wanted to call attention to the inadequacy of thinking only in terms of the ethics of technologies, that is, the responsibilities of individual engineers and users of technologies. Any ethics of technology by itself will tend to be weak if not impotent. Ethics deserves to be supplemented with policy.

ZZ: So the policy turn is another aspect of the applied turn in ethics?

CM: To some extent. But the applied turn in ethics too often remains focused on individual responsibilities. Stephen Toulmin's famous article "How Medicine Saved the Life of Ethics" (1982) argued that the ethical conundrums of high-tech medicine brought ethics out of the linguistic analysis clouds to deal with substantive issues in the clinic. Applied ethical analyses nevertheless tended to remain relevant to individual cases and patients. Casuistry is not enough. Biomedical ethics requires the supplemental formulation of biomedical ethics policies—and ultimately the adoption of laws. An article with Erik Fisher (2012) made an effort to reflect more, at least descriptively, on this topic.

Let me also note that the notion of a policy turn has played a role in environmental ethics and philosophy. Let me quote Frodeman here:

> A policy turn in environmental philosophy means a shift from philosophers writing philosophy essays for other philosophers to doing interdisciplinary research and working on projects with public agencies, policy makers, and the private sector. Despite some steps in this direction, a policy turn remains largely unrealized within the community of environmental philosophers. Completing this shift can contribute to better decision making, help discover new areas for philosophic investigation at the intersection of philosophy and policy, and identify new employment prospects for philosophy graduates. (Frodeman, 2006, p. 3)

Interestingly, at the annual meeting of the Society for Social Studies of Science (4S) at Pasadena in 2006, in a panel on laboratory studies, Bruno Latour made a plea for more policy relevant STS.

The problem with laboratory studies, from my perspective, is that they were not truly interdisciplinary. They took STS analysis and philosophy into the laboratory, but only with their own STS and philosophical questions. This is a one-sided or one-way interdisciplinarity. In a more robust, two-way interdisciplinarity, philosophy would be more deeply challenged by the science or technology with which it was interacting.

Subsequently, Frodeman and his colleagues have used the term "field philosophy" to name what I consider another aspect of the policy turn. For that same 4S meeting, I participated with Frodeman in proposing a panel session on "Field Experiments in Interdisciplinarity: New Directions: Science, Humanities, Policy" explicitly linking the two.

ZZ: In one of your articles from this time period (Mitcham, 2009) you connected policy with what you called "post-engineering." This is another strange term that people find it difficult to understand. What do you mean by "post-engineering"?

CM: I argued there that engineers in their concerns with professional ethics had, even if they did not realize it, already become involved with policy, insofar as they expressed dissatisfactions with codes of ethics that placed all responsibilities on individual engineers and did not consider the social contexts of engineering practice. To propose accreditation requirements for teaching engineering ethics and by trying to provide professional organizational support for engineers who make ethical judgments at odds with corporate or government employers, they were doing policy work. Accreditation requirements are policies, which aspire to be based on empirical knowledge about what is required for good engineering practice and what works. The whole field of engineering education research seeks to replace intuitions about how best to teach engineering, including teaching engineering ethics.

But then I went further to suggest that an increased sense of the importance of professional ethics in the English-speaking American engineering context implies a new form of engineering that I termed "post-engineering." This was meant to be both playful and provocative. Classical English-speaking engineering did not give ethics a very significant role in the profession. It was just assumed that engineering is good, always good. Insofar as ethics becomes an important and also contested element in engineering life, a new form of engineering is emerging. "Meta-engineering" might have been a better term.

The basic point is that as our world becomes more and more engineered and dependent on engineering, engineering ceases to be just one part of culture. It takes over. Let me quote again a passage from an insightful memoir by a former MIT Dean of Undergraduate Education, Rosalind Williams. (I quoted it in the article you referenced.) Williams describes what she calls "expansive disintegration" of engineering.

> There is no "end of engineering" in the sense that it is disappearing. If anything, engineering-like activities are expanding. What is disappearing is engineering as a coherent and independent profession that is defined by well-understood relationships with industrial and other social organizations, with the material world, and with guiding principles such as functionality .... Engineering emerged in a world in which its mission was the control of non-human nature and in which that mission was defined by strong institutional authorities.

> Now it exists in a hybrid world in which there is no longer a clear boundary between autonomous, non-human nature and human generated processes. (Williams, 2002, p. 31)

In such a world there is also no longer any clear boundary between engineering and non-engineering. Engineering has like modernity gone beyond itself into a post- or meta-state.

10.6 The Empirical Turn in the Philosophy of Technology

ZZ: Your remarks on the applied turn and policy lead naturally such another question: What about the "empirical turn" in the philosophy of technology? Is there a relationship between these other "turns" and the empirical turn?

CM: I think so, but proponents of the empirical turn might disagree. The empirical turn in the philosophy of technology was constructed as part of an argument for change. It had an ideological function. I say this not as a criticism but simply to recognize that in some ways it was as much rhetorical as philosophical, an effort to marshal cultural capital. The same may be said for the terms "applied turn" and "policy turn."

The polysemy of "empirical turn" is not always appreciated. It originated, as you know, with Dutch philosophers in the late 1990s: especially with Hans Achterhuis at Twente University and with Peter Kroes at TU Delft, each with slightly different interests. Fortuitously, I had a fellowship to the Netherlands in spring 1998, visiting with René von Schomberg at Tilburg University, with Achterhuis at Twente, and with Kroes at Delft. This was when I first learned of what I have come to think of as the "dual (dueling) empirical turns."

The term first appeared in the title of a "Workshop on the Empirical Turn in the Philosophy of Technology" held at TU Delft, April 1998. In his programmatic introduction, Kroes argued that because philosophy of technology had been ineffectual, it needed to undergo a three-fold reorientation: (1) from user phase to the design, development, and production phase; (2) from global to local analysis; and (3) toward use of empirical case studies. A proceedings volume from that workshop was edited by Kroes and Anthonie Meijers and published as a theme issue of *Research in Philosophy and Technology* (vol. 20, 2000). Part four of this volume was on "Ethics and the Empirical Turn" and included "The Ethics of Engineers: From Occupational Role Responsibility to Public Co-Responsibility," which I co-authored with von Schomberg. That article was a kind of exercise in applied ethics, although we did not call it that.

When Achterhuis used the term, however, it was to subtitle an English translation of his *Van Stoommachine tot cyborg, denken over techniek in de nieuwe wereld* (1997, "From steam engine to cyborg, thinking about technology in the new world"): *American Philosophy of Technology: The Empirical Turn* (2001). Instead of proposing a programmatic change that should take place in the future, Achterhuis claimed to be describing a turn that had already taken place among a group of Americans:

Albert Borgmann, Hubert Dreyfus, Andrew Feenberg, Donna Haraway, Don Ihde, and Langdon Winner.

To emphasize his point, Achterhuis distinguished between what he called the "classical philosophy of technology" of such figures as Lewis Mumford, Martin Heidegger, Hans Jonas, Jacques Ellul, Hannah Arendt, and Ivan Illich, who "occupied themselves more with the historical and transcendental conditions that made modern technology possible than with the real changes accompanying the development of technological culture." In so doing, they overlooked "the manifold ways in which technology manifest itself" (2001, p. 3). Achterhuis thus expressed appreciation for the achievements of the classical philosophers and their discoveries that technology is not just applied science and more a form of life, a "system" (Ellul) or "Megamachine" (Mumford). "It is the great merit of the classical philosophy of technology to have perceived this as amounting to a new and radical turning point in the history of humanity" (p. 4). Kroes expresses much less appreciation for pre-empirical turn insights.

It is remarkable, too, that there was only one common contributor to both the Kroes-Meijers and the Achterhuis volumes: Philip Brey from Twente co-authored with engineer Clive Dym an examination of engineering design languages for Kroes-Meijers and single authored a study on Hubert Dreyfus for Achterhuis. None of the philosophers profiled in Achterhuis are given any significant treatment in Kroes. There were clearly something like two empirical turns being advanced here.

Yet the three features that Achterhuis proposed as key to the empirical turn, which he says has already taken place, do in fact echo two of the three programmatic proposals from Kroes. Achterhuis describes American philosophers of technology as (1) opening the black box of the social construction of artifacts (another version of Kroes #1); (2) looking not at technology *tout court* but at the multiplicity of technologies (similar to Kroes #2); and (3) appreciating the co-evolution of technologies and societies. Achterhuis third key characteristic actually seems to have something in common with the achievements of classic philosophers of technology.

ZZ: You think that the empirical turn for Kroes and Meijers was not the same as the empirical turn of Achterhuis, and that Achterhuis is more appreciative of classical philosophy of technology.

CM: That's correct. Another more hidden indication of this is that during the same spring of 1998, following publication of the Dutch version of what became in English his empirical turn book, he hosted a "living room conversation" with his old friend Ivan Illich on the problem of how to live in the radically new techno-human condition. I would say that Achterhuis, unlike Kroes, did not simply turn away from classic philosophy but wanted to complement it.

At the same time, I do not fully agree with Achterhuis's characterization of either classical philosophical of technology or his empirical turn Americans. Classical philosopher of technology Mumford, for example, did extensive work on how particulars such as the mechanical clock manifested modern technology. Empirical turn philosopher Borgmann thinks not just about particular technologies but tries to elucidate the "device paradigm" present throughout modern technology as a whole. Additionally, I think the interpretation of pre-empirical turn philosophers as focusing on "transcendental conditions" is inadequate. Ellul, and Illich, at least, were as much if not more focused on the cultural consequences of modern technology as on any transcendental preconditions.

ZZ: With regard to Achterhuis and Twente University, I wonder if there is any relationship between the philosophy of technology of Peter-Paul Verbeek, who is a professor there, and his mediation theory, which is becoming well known in China.

CM: This is a good question, since Verbeek has in effect established another Dutch school in the philosophy of technology. Verbeek did his PhD with Achterhuis, although he has since worked more closely with Don Ihde in the promotion of Ihde's postphenomenological program. Verbeek's book *What Things Do* (2005) sets the stage for his mediation theory by drawing on Achterhuis's characterization of classical philosophy of technology to criticize in detail the work of Karl Jaspers and Martin Heidegger. Then, as you note, he makes a programmatic argument for what he calls a mediation theory of technology. In this regard he does some interesting research, but again I think "mediation" is more a rhetorical branding than a significantly new trajectory in thinking.

The postphenomenology brand of philosophy of technology has definitely flourished under Ihde's leadership. Ihde taught a generation of younger philosophers, publishes prolifically, and been tireless in conference presentations. (Incidentally, the influence of Ihde in China has recently been comprehensively reviewed by Zhang Kang, 2020.)

ZZ: Does the TU Delft empirical turn school acknowledge the Twente empirical turn school or vice versa?

CM: To some degree. But, as indicated, the Twente use of "empirical turn" was not programmatic in the same way as the Delft empirical turn. When Verbeek puts forth a program for research, he either allies himself with Ihde's postphenomenology (phenomenology of instrumentalization instead of intentionality) or his own term "mediation theory." And just as the Delft empirical turn philosophers largely ignore Achterhuis, Verbeek, and Ihde, so the postphenomenologists and mediation theory philosophers do not pay much attention to work by members of the Delft school.

ZZ: So what has become of the Delft empirical turn over the course of now close to 20 years?

CM: The Delft empirical turn school has been very successful in creating a strong analytic and argumentative community of discourse in the philosophy of technology, something that had not previously existed. Classical philosophy of technology, as found in José Ortega y Gasset, Martin Heidegger, and Jacques Ellul exhibits philosophy in a different mode: philosophy as an invitation to notice or attend to something that had been missed (philosophy as a deconstruction of the techno-lifeworld, to repurpose a prominent methodological appellation). It was not trying to establish an academic argumentative community of discourse, a research program in which members could compete to refine concepts and clarify problems (as is the case in analytic philosophy of science). Ortega, Heidegger, and Ellul wanted to function more like poets who help us see and experience realities that we have otherwise overlooked: the basic reality being how modern technology has introduced a rupture or tear or mutation in the continuity of the human lifeworld. The Delft school is not particularly interested in philosophy in this big picture sense.

Confirming these points, an edited volume by Maarten Franssen et al. on *Philosophy of Technology after the Empirical Turn* (2016), put together (in part) in honor of Kroes' retirement, makes no mention of Achterhuis. There are also little to no references to classical philosophers of technology identified as such by Achterhuis. Interestingly, however, Adam Briggle (another colleague of Frodeman and mine) contributed an article on "The Policy Turn in the Philosophy of Technology."

> The *After the Empirical Turn* collection provides the best single introduction to the scope and achievements of the Delft empirical turn program and the current state of what I would call, more generally, analytic philosophy of technology. As the editors write in their introduction, the empirical turn succeeded in steering the philosophical study of technology away from broad abstract reflections on technology as a general phenomenon toward addressing philosophical problems that can be directly related to "the way technology works" or to "technology in the making". In so doing, it focused primarily on the work of engineers. Accordingly, one of its principal messages was to urge a shift of focus to the design of technical artifacts, rather than their later career as constituents of practices. (Franssen et al., 2016, p. 2)

The 17 contributions by 20 authors mostly see "the empirical turn [as] a fruitful signpost to follow for the near future of the philosophy of technology" (p. 8).

But returning to your question about relationships between the applied, policy, and empirical turns, the empirical turn can be described as aspiring to deepen the applied turn and, insofar as possible, incorporate the policy turn—and to do so in a more empirically robust and analytically rigorous manner than in the past. The substance of a policy turn (if not the term) has been picked up by the Twente school more than the Delft school. Philip Brey, who succeeded Achterhuis as Chair of the Twente Philosophy Department, uses the term "anticipatory technology ethics" with an intent related to providing policy advice (see, e.g., Brey, 2012). But the policy turn has perhaps been more characteristic of STS studies than of philosophical research. The applied and empirical turns are addressed to individual decision making, while the policy turn is concerned with governmental decision making.

Finally, just to emphasize again what I've already said, the program associated with the empirical turn in the philosophy of technology is currently the most robust discourse community in philosophy and technology studies. It has been the most successful in turning philosophy *and* technology studies into philosophy *of* technology in a way that garners respect in the larger academic philosophical world. It will be interesting to see to what extent Chinese philosophers of engineering and technology become contributors to this community of argumentative discourse or draw from it in the future.

ZZ: So you think the empirical turn is something that all philosophers of technology should adopt, that it has definitely replaced classic philosophy of technology.

CM: Not at all. In philosophy, one approach or school of thinking is never able to definitely replace all others. Recently an Italian philosopher, Agostino Cera, put forth a provocative criticism of the empirical turn. Cera (2020) argues for distinguishing two moments in the empirical turn: A legitimate turning away from Heideggerian mysticism of technology as *Ereignis* (event) and a subsequent illegitimate rejection of all ontological thinking concerning technology. Empirical turn philosophers often accuse classic philosophers of exhibiting *techno-phobia*. According to Cera, however, their own *techno-philia* has become an *onto-phobia* or an "a priori disinterest in any ontological implication of technology." The empirical turn turns questions concerning technology into problems concerning technology or what he also characterizes as the "engineerization" of thinking. Cera invites us to notice something that we have been overlooking about the empirical turn and thus further invites a renewal of traditions occluded by a too ready rejection of classic philosophy of technology in the name of effectiveness.

10.7 Appreciating Chinese Philosophy

ZZ: A number of times now you have mentioned a growing Sinophilia and appreciation for Chinese culture, history, and philosophy. Could you say more about your approach to Chinese philosophy, especially traditional Chinese philosophy?

CM: Ever since my 2011 visit to Qufu and Taishan—and in subsequent years I climbed Taishan a second time as well as Huangshan, while visiting numerous Daoist, Confucian, and Buddhist temples—I have been increasingly attracted to the way Chinese culture (文化) integrates poetry and philosophy. Tang and Song poetry is unlike anything that I know in the West. The novels of Zhang Ailing and Wang Anyi strike me as stunning expressions of what, for want of a better term, I consider a non-Abrahamic sensibility. In the Abrahamic religions of Judaism, Christianity, and Islam there exists just below the surface (if not openly exposed) a kind of fanaticism for justice that often makes justice impossible to negotiate. Wang Anyi's长恨歌, which invokes the classic Tang tale of Yang Guifei and is translated as "Song of Everlasting Sorrow," is a narrative of regret that nevertheless remains a combined lament and lyric celebration of life within the limits of a human condition devoid of supernatural redemption.

ZZ: How does this relate to Chinese philosophy?

CM: I take it as an expression of what Li Zehou (whom I have been privileged to meet and already mentioned) calls the one-world ontology of Chinese civilization. Two English language books by Li—*The Path of Beauty* (1988) and *The Chinese Aesthetic Tradition* (2010)—provide a foundation for appreciating not just Chinese literature but traditional Chinese art and architecture (as I've learned about it through Liang Sicheng and Lin Huiyin).

However, my initial introduction to Chinese philosophy came by way of Joseph Needham's well known exploration of how the development of science and technology took place so differently in China and in the West. My other initial tutor was Feng Youlan. I'm told now that Feng's *A Short History of Chinese Philosophy* (1948) is not a high quality translation, but it remains a masterful overview; and his autobiography, *The Hall of Three Pines* (1999), is a moving narrative. His return to

China in 1949 out of a desire to contribute to the New China and the way he honorably suffered persecution are both admirable. To me, Feng was truly noble, a twentieth century *junzi* 君子.

May I venture to say something a little more general about the new interest in Chinese philosophy that has emerged in the West?

ZZ: Yes, of course.

CM: In my efforts to study the Chinese philosophical tradition I have naturally been forced to rely on English translations and English language interpretations. For a few years now I have been studying the language, which fascinates me, but I am just too old to ever learn it. As I tell people, I am not learning Chinese; I am just learning *about* Chinese. Still, learning even the little that I have is helpful. I plan to continue as best I can.

In the background of any American approach to Chinese philosophy and culture is Ezra Pound (1885–1972), the greatest American poet of the twentieth century. As critic Hugh Kenner (1971) says, twentieth century English poetry was "The Pound Era." In a review of *Cathay* (1915), a Pound interpretation of selected classical Chinese lyrics, T.S. Eliot described him as "the inventor of Chinese poetry for our time." Pound had discovered Chinese poetry 2 years earlier through Ernest Fenollosa (1853–1908), when his widow gifted Pound the papers of the American Orientalist, Japanese art historian, and practicing Buddhist who taught for much of his life in Tokyo. Fenollosa's *The Chinese Written Character as a Medium for Poetry*, which Pound edited and published in 1920, was an eccentric but influential, misreading of the Chinese script. But for Pound it pointed the way toward a revival of English language poetry. Taking Chinese poetry as a new model would free English poetry from worn out sentiments and dog-eared forms to focus on language condensed and intensified in images. Additionally, Pound became so smitten by Confucianism—or his idiosyncratic version of Confucianism—that he saw it as the proper foundation for a reform of Western political affairs after the catastrophe of World War I. (One irony is that at the same time as American poets were looking to China for inspiration, Chinese poets and writers in the New Culture Movement such as Lu Xun were looking to the West for a revival of Chinese literature and political affairs.)

Sinologists, as one might expect, have been dismissive of Pound. Scholarly discussion of Pound and China has instead existed largely in its own bubble of poets and critics (see, e.g., Qian, 2003). Yet I think there is something philosophical to be learned from Pound.

ZZ: I'm not sure what you mean by Sinologists. Is Sinology the same as China Studies?

CM: Not exactly. Sinology originated in Europe (mostly in France, Germany, and England) in the early 1800s but can be trace back to Matteo Ricci and early European encounters with the Ming court. It is companion to classical philology—philology applied to Chinese—and thus entails detailed study of classical Chinese language, literature, and philosophy (just as classical philology in Europe entailed the study of the classical Greek and Latin languages, literature, and philosophy). Early in his career, Friedrich Nietzsche was Professor of Classical Philology at the University of Basel. Sinology is a discipline. China Studies, by contrast, is an interdisciplinary field focused of learning that draws on Sinology along with history, politics, anthropology, sociology, economics, and more, to engage both past and present.

China area studies emerged originally in an American strategic effort to understand China for diplomatic and military reasons. The U.S. State Department and Pentagon were the initial funders of a first generation cohort of scholars who learned Chinese and were sent to Mainland China but very quickly had to retreat to Hong Kong and Taiwan. During the 1970s China Studies distanced itself from American strategic interests but tensions remain between it as the discipline of Sinology. The Australian public intellectual sinologist Geremie Barmé (2005), for instance, argues for a "New Sinology" that resists incorporation into China Studies.

For myself, however, I find both the political and anthropological contributions of China Studies informative and complementary. Some of the "grand old men" of China Studies are still worth reading, if only for their revealing display of Western biases: the "China watchers" John King Fairbank (1907–1991) who did diplomatic history (and influenced Henry Kissinger, the adviser who facilitated President Nixon's opening to China), and Benjamin Schwartz (1916–1999), another political scientist; and William Theodore de Bary (1919–2017), who opened the Columbia University humanities core to non-Western literature. De Bary had more influence on philosophers than Fairbank and Schwartz, who were based at Harvard. I've read selections from all three (as well as Kissinger's *On China*, a typical *Realpolitik* analysis).

However, contributions to China Studies from humanities scholars (among whom De Bary was a pioneer) are more fruitful. Although I have done only the most cursory reading in this area, allow me to reference three very different examples that help inform my philosophical studies. Social historian Francesca Bray, who contributed the volume on agriculture to the Needham *Science and Civilization in China* project, has two studies of technology and gender in China (1997, 2013) that open up new worlds. The rich descriptions of embodied life in post-Mao China by medical anthropologist Judith Farquhar (Farquhar, 2002; Farquhar & Zhang, 2012) are exemplary invitations to appreciation of quotidian practices of *yangsheng* 养生 (nurturing life). Sociologist Joel Andreas's the *Rise of the Red Engineers* (2009) provides a specialized analysis of this technical profession at one particularly crucial point in time.

In the philosophical component of China Studies we can notice two trajectories: In the West, coming from the outside, through new discoveries and appreciations of Chinese culture as containing a genuinely philosophical tradition; in China, taking place on the inside, through renewals and advancements of multiple Chinese philosophical traditions. Since becoming a visiting professor at Renmin University I have been struggling to orient myself in these two constellations of philosophy, as a necessary engagement for allowing Chinese philosophy to teach me—teach me, especially, with regard to the challenges embedded in modern engineering and technology.

ZZ: It might be helpful for you to elaborate more on this distinction between two trajectories for renewing traditional Chinese philosophy. As a historian and philosopher of technology, this is not a distinction that is obvious to me.

CM: But I'm sure as a historian you are aware of some similarities and differences in how Chinese and Western scholars approach the history of science and technology in China—or even just the history of China. Paul Cohen (2010), an American historian of China, has noted how Western historians have tended to adopt one of three theories when interpreting the modern history of China: impact-response, modernization, or imperialism. He argues for replacement of all three by a "China-centered history" that pays more attention to the ways that China and Chinese historians have interpreted their history. Similar tensions between inside and outside exist in the revivals of Chinese philosophy.

Within the Chinese philosophical community there were, first, Hu Shih (1891–1962) and Feng Youlan (1895–1990), who attempted at once to bridge East and West and to revive Chinese philosophy. They were criticized by Xiong Shili (1885–1968), Liang Shuming (1893–1988), Tang Junyi (1909–1978), and Mou Zongsan (1909–1995) for too much subservience to Western thinking and failure to acknowledge the true genius of Chinese. (This criticism seems to me to apply less to Feng than to Hu.) Tang and Mou, although born in the mainland, moved to Hong Kong after the founding of the New China. Mou's work is slowly being translated into English. Another leading figure who emerged in the cultural fever of the 1980s is Confucian Marxist Li Zehou (b. 1935), whose studies of Chinese aesthetics have been particularly influential on me. But for an English reader, Chung-Ying Cheng and Nicholas Bunnin's edited volume on *Contemporary Chinese Philosophy* (2002) is perhaps the best introduction to this community of discourse. A complementary volume is Guo Qiyong's *Studies on Contemporary Chinese Philosophy (1949–2009)* (2018).

With regard to the outside, which is where I have to begin, I would distinguish three different modestly overlapping schools centering around (1) Angus C. Graham (1919–1991) at the School of Oriental and African Studies in London, (2) Donald J. Munro (b. 1931) at the University of Michigan, and (3) Roger Ames (b. 1947 in Toronto) at the University of Hawaii.

Ames is incredibly prolific, but appears to have had fewer students than Munro (whose mentees include Robert Eno, Stephen Angle, Chan Hansen, Ni Peimin, An Yanming, and more). Both Ames and Munro identify as Confucians, or rather as something we might call American Confucians; both studied with and revere the New Confucian philosopher Tang Junyi.

Although no longer considered a necessary place to begin, Munro's *The Concept of Man in Early China* (1969) was a pioneering analysis of *xing* 性and *de* 德. This was a creative philosophical step beyond the kind of broadening cultural education pedagogy promoted by de Bary. Munro was also unique in his research into Marxist philosophy with *The Concept of Man in Contemporary China* (1977). The best general introduction to his work is a *Festschrift* edited by Yanming An and Brian Bruya (2019). An's own *The Idea of Cheng (Sincerity/Reality) in the History of Chinese Philosophy* (2005), which brings the methods of Dilthey's hermeneutics to bear on one of the fundamental ideas in Chinese thinking to yield deeply original results, is another testimony to the high quality of Munro's mentorship. It was a meeting with An Yanming at Dalian University of Technology in 2007 that first opened my eyes to the work of Munro.

Ames's collaborative trilogy (Hall & Ames, 1987, 1995, 1998) together with his *Confucian Role Ethics* (2011) constitutes an extensive argument for Confucian philosophy as an alternative and response to the weaknesses of modern Western philosophy. As editor of the journal *Philosophy East and West* for more than 30 years (1986–2017) Ames exercised a formative influence.

However, to my mind, the most important Sinologist, philosopher, translator, and poet is the British scholar Angus C. Graham. His *Later Mohist Logic, Ethics and Science* (1978) and *Disputers of the Tao* (1989) are recognized classics of Western discovery and interpretation. The former is a fundamental revalorization of the significance of Mozi as philosopher; the latter offers a new history of classical Chinese philosophy as one not just of competing poetic ideals but of argument and counter-argument. *Reason and Spontaneity* (1985) is an ambitious metaphysics, epistemology, and ethics that integrates Chinese and Anglo-American philosophy. Additionally Graham has published studies and translations of Neo-Confucianism, legalism, Zhuangzi, and Tang poetry. But one special indication of his stature is the publication of two *Festscrhtiften*.

The first, *Chinese Texts and Philosophical Contexts* (1991), edited by Henry Rosemont Jr., had 10 engaging an incomplete range of Graham's achievements: on philology by D.C. Lau and Edwin Pulleyblank; on translation and interpretation of classical texts by Christoph Harbsmeier, John Major, Harold Roth, David Nivison, and Roger Ames; and on philosophical issues by Chad Hansen, Herbert Fingarette, and Henry Rosemont Jr.

The second, edited by Carine Defoort and Roger Ames (2018) is a unique posthumous *Festscrift* with 12 contributions (only three authors repeated from the previous volume) indicating the continuing fecundity of Graham's thought for a new generation of scholars: on Graham's extensive contributions to *Zhuangzi* studies by Esther Klein, Liu Xiaogan, Harold Roth, and Michael Nylan; on translation and interpretation by Henry Rosemont Jr. and Robert Gassmann; on the *Mengzi* by Jane Geaney, Defoort, and Ames; and on Graham's metaphysics of spontaneity by Lisa Raphals, Paul Kjellberg, and Chris Fraser.

One distinction I've noticed is that between scholars who teach in some type of Chinese Studies programs (e.g., de Bary) versus those in philosophy departments (Munro and Ames). In some ways I find myself as much attracted to a new generation of scholars of Chinese literature such as David Der-wei Wang at Harvard and Rey Chow at Duke as to philosophers—which fits with my sense of the distinctive role literature has played in Chinese society. There is nothing in the West like the thousands of years of continuity in cultural history that provides a kind of stability for Chinese identity or the literati ideal that was for centuries the foundation for government service. Wang's (2017) concept of "worlding literary China" and Chow's (1998) of "ethics after idealism" introduce counterpoints for bridging the inside and outside dialectic of recovery and discovery discourse.

Then of course there are also schools of French scholars and German scholars, about whom I know almost nothing. But I have gone on way too long. I hope you can forgive this excessively personal literature review.

ZZ: No, I think it is a useful perspective by an American scholar on Chinese philosophy, and for what it suggests about your own engagement with Chinese philosophy.

10.8 Chinese Philosophy of Technology for the West

ZZ: I would like now to conclude by returning to Chinese philosophy of technology. I know you have supported publication of an English translation for one of Professor Li Bocong's books. Could you say something about why you see this as important?

CM: In 1992, when I first met him, Li Bocong gave me a copy of his book扁鹊和扁鹊學派研究 *Bianque he Bianque xuepai yanjiu* (1990), which you have pointed out to me is highly regarded by Western historians of traditional Chinese medicine. But my first visit to China was something of a blur, and it took me years to recognize the importance of Li's work. It was not until 2 years later, in 1994, when he came to the United States as a visiting scholar, that I became able to pick him out of a crowd—and began to appreciate his scholarship. From the beginning, he knew my work much better than I knew his, just as even now I think he understands Western philosophy better than I understand Chinese philosophy.

Li Bocong became the single person from whom I have learned the most over more than 25 years of inquiry into Chinese philosophy of technology. This education includes, of course, learning from some of his leading students such as you.

I might add that there is more in appreciation of Li Bocong's work in my *Steps toward a Philosophy of Engineering* (2020). And I have been asked to contribute a foreword to the English publication of Li's forthcoming *An Introduction to the Philosophy of Engineering*.

ZZ: From your perspective, what is the most important thing you would say you have learned from Professor Li?

CM: Beyond being edified by his rigorous life as an interdisciplinary philosopher engaged with the international scholarly community, I would list three things: The central place of engineering in the philosophy of technology, the inherent attractiveness of the what I call heroic engineering, and the need to distinguish Chinese工程 *gongcheng* and English "engineering." In regard to all these points, I would add that I am still learning.

ZZ: Are there other Chinese philosophers of technology who have been important to your work?

CM: There are at least three others: Professor Chen Fan at Northeastern University (NEU) along with some of his colleagues and students, Professor Wang Qian at Dalian University of Technology (DUT) and his colleagues and students, and Professor Liu Yongmou at Renmin University of China (RUC) along with the whole Renda Philosophy of Science and Technology Department.

ZZ: Could you say more about each of these?

CM: The association with Chen Fan also goes back to my 1992 lectures, which he attended. He had just finished his PhD that spring and gave me a copy of his dissertation on技术社会引论:种对技术的社会学研究 *Jishu shehui yinlun: Zhong dui jishu de shehui xur yanjiu*, which he translated as "Socialization of Technology: A Sociological Study of Technology." I thought his contrast between technological industrialization and socialization, in the second case trying to make sure technology is for social benefit, was a useful concept related to STS studies in the United States. Later that year Chen wrote telling me that his academic home, Northeastern Institute of Technology, had been renamed Northeastern University, established a College of Liberal Arts and Law, and created a new Institute of Technology and Society for which he would serve as director.

Chen Fan became a leader in promoting a broad, STS approach to the philosophy of technology with which I am sympathetic. He also became a leader for inviting American philosophers of technology to China and in translating these philosophers into Chinese. I visited Northeastern University on my travel to China in 1998 and a number of times since as well as hosted two of Chen Fan's students as predocs: Chen Jia (in 2009–2010) and Yin Wenjuan (in 2012). And as you know, Chen Fan supervised the 2008 Chinese translation of *Thinking through Technology*.

Furthermore, Chen Fan played a pivotal role in bringing the Society for Philosophy and Technology (SPT) to Northeastern University for its 19th international conference July 2–5, 2015, hosted by the NEU Research Center for Philosophy of Science and Technology and the Chinese Society for Dialectics of Nature/Philosophy of Nature, Science and Technology.

The most important thing I learned from Chen Fan, however, concerned the existence and unique achievements of the "northeastern school" of the philosophy of technology. I want to acknowledge you also as one of my teachers in this regard.

As I look back over my years visiting China I keep discovering ways I missed recognizing important things. As just one more example, the Gao Dasheng and Zou Tsing (1989) article, which I mentioned earlier, called attention to work by both Chen Changshu and Chen Fan. But it was really not until your interview that, in looking back through my files, I realized this.

Along this line of making up for things missed, I cannot help but mention my great good fortune at the 2015 SPT conference to be able to meet and interview (through the good offices of Yin Wenjuan and Cao Dongming) Yuan Deyu, one of Chen Changshu's most important collaborators and an important hidden philosopher in his own right. (An English version of this interview was published in Mitcham et al. (2018).)

ZZ: With regard to the northeastern school in the philosophy of technology, another member of that school was Liu Zeyuan at DUT.

CM: That's right, and this is the perfect segue to saying something about my associations with DUT. My first visit there took place in 2007 when Professor Wang Qian invited me to participate in an "International Conference on Science and

Technology Ethics and Business Ethics" (July 16–18). There I discovered another center of philosophy of technology teaching and research led at that time by Dean Hong Xiaowei and including Liu Zeyuan and Wang Guoyu. Liu Zeyun's contributions to the international scientometrics research program and Wang Guoyu's work with the philosophy of technology tradition in Germany particularly surprised me. Wang Qian, Liu Zeyuan, Wang Guoyu, and what I have sometimes referred to as the "DUT school" all deepened my knowledge of the state of philosophy and technology studies in China.

ZZ: Is there anything more particular about this "DUT school" that strikes you as important for the western philosophers of technology?

CM: Wang Qian's research on Daoism and process philosophy has challenged my thinking, even though I haven't incorporated it much into my own work. I remember one afternoon on a subsequent visit benefiting from a private seminar in which Wang Qian shared a close reading of the Butcher Ding story from the *Zhuangzi*. His student Zhu Qin, who acted as interpreter, became my first Chinese predoc, and now is an Assistant Professor colleague at Colorado School of Mines. Before being appointed visiting professor at Renmin University in 2015, I had spent more extended periods of time at DUT than any other institution.

The two of my books that appeared only or first in Chinese were collaborations with DUT: 同行评议、研究诚信与科学治理:实践、理论与当代议题 (co-authored and co-edited with Robert Frodeman and J. Britt Holbrook, 2012) and工程与哲学:历史的、哲学的和批判的视角 (2013). The second one, much revised and expanded, became *Steps toward a Philosophy of Engineering*.

I should mention as well that Yan Ping, under the direction of Professor Wang Qian, has done an insightful study of my thinking about engineering ethics. The perspective that another person can provide on one's work can be quite informative.

ZZ: I guess it is now time for you to say something about your engagements at Renmin University.

CM: Being appointed a visiting professor at Renmin University of China since 2015—I actually gave my first short course there in December 2013—has created a new academic home for my old age. I am most grateful to Professor Liu Yongmou and the community of scholars and students in the Department of Philosophy of Science and Technology of the School of Philosophy for immeasurably broadening and deepening my knowledge of China, of Chinese history, culture, and philosophy—including the Chinese language (since I was able to sit in on a Chinese course for international students). Without minimal appreciation of the language and culture, it is difficult to grasp some of the distinctively Chinese perspectives on technology from which we in the West might learn. At Renda I have slowly learned something of the larger contexts of my previous engagements with the University of Chinese Academy of Sciences, Northeastern University, and Dalian University of Technology and to qualify some of my distorting western assumptions.

ZZ: What would you say are your most "distorting western assumptions"?

CM: The single most important one, with which I first came to China, was that the primary relationship between China and the West was that China needed to learn from the West. I no longer think that is the case. In fact, I would almost reverse it and say that the West, or at least America, has more to learn from China than China has to learn from America. Yet there is a paradox here, because I think what Americans may need to learn most from China is the way China is learning from the West.

ZZ: Are there particular scholars at Renda who you see as important contributors to the philosophy of technology, especially for philosophy of technology in the West?

CM: The scholar with whom I have collaborated most is Professor Liu Yongmou, whose ideas and arguments about technocracy or technoscientifc governance deserve wider dissemination in the West, especially as the American public becomes increasingly distrustful of science. As already mentioned, I am deeply indebted to Liu Yongmou for opening the door for me at Renmin University, and (as with Professor Li Bocong) I would like to help facilitate the publication of more of his work in English.

I have also learned from his professor and senior member of the department, Liu Dachun, to whom he introduced me. In the process of interviewing Professor Liu Dachun I discovered a scholar of depth and grace. His 科学之审度 *Kexue zhi shen duo* [Reconsiderations of Science] collects 28 papers from the 1980s to the 2010s on philosophy of science, philosophy of technology, STS, philosophical reform in China, and the humanities and social sciences more generally. What was most illuminating for me is an intellectual autobiography that tells the story of how, despite adversities during the Cultural Revolution, he was able to pursue philosophical studies and become a leading contributor to philosophy at Renmin University and in the Chinese intellectual community as a whole. Cheng Changshu, among many others, was influenced by his work.

I would very much like to read two other of Liu Dachun's books: an edited study of Marxism in relationship to contemporary ideas about STS (2017) and a reflection on the Chinese appropriation of Western knowledge (2019). In the former volume, the contribution of Wang Boulu strike me as especially valuable. Unfortunately, because of my lack of Chinese I am limited in my ability to appreciate the substance of these volumes. I am dependent on the help of students who have tutored me about their content.

Other Renda faculty from whose work and collegial friendship I have especially profited include Liu Jinyang, Wang Xiaowei, and Teng Fei. And I should give a lot of credit to many of my students, especially my assistants, for their hospitality, tolerance of my lack of Chinese, and dedication to helping me better understand Chinese perspectives on philosophy and the world. Without the enthusiasm and dedication of these students, my intellectual life in China would not be nearly so rich.

ZZ: Are there other persons or institutions besides these that you think have particular perspectives that might be of benefit to western philosophy of technology?

CM: There are certainly many others who model the practices of hospitality and desires to learn that I see as one of the most important things China models for the West. I have visited more Chinese universities than I can quickly remember. Tshinghua University, beginning in 1992, I have visited a number of times, certainly stands out. Most recently Professor Wu Guosheng, who was previously at Beijing

University, has been very hospitable and challenged me to think not just in terms of philosophy but of Chinese culture (文化) more generally. Huang Xiaowei, a doctoral student from Tsinghua who did a predoc year with me in Colorado, has educated me about Chinese work in engineering ethics. Relationships with the Chinese Academy of Social Science also go back to the early 1990s, and most recently I have been learning about ethical and political issues engendered by artificial intelligence from Professor Duan Weiwen. I've had discussions about science policy with professors at Beihang University.

Outside of Beijing, I should also mention collaborations with Professor Wang Guoyu now at Fudan University on nanotechnology and Professor Xue Guibo at Nanjing Forestry University on environmental issues. A short visit to South China University of Technology in Guangzhou introduced me to another philosophy of technology program led by Professor Wu Guolin.

10.9 Conclusion

CM: Maybe this is enough for the present. I would like to conclude by thanking you for the stimulation of this interview. Your questions have encouraged me to become more conscious of my relationships with China.

ZZ: It is my pleasure. It has stimulated me as well. But I wonder here at the conclusion what words you might have for young scholars in China who are working on philosophy of technology or engineering studies?

CM: I am uncomfortable making any strong recommendations. I think my relationship is different from many other philosophers who go to China taking their well-developed approaches and theories, for example in the cases of Don Ihde, Andrew Feenberg, or Bruno Latour. At least since the early 2000s, I have a sense that Chinese history and culture have at least as much to teach me as I have to teach Chinese students and scholars. Earlier in my career I tried to call to the attention of English speaking Americans some of the underappreciated critical reflections on technology by French, German, and Spanish philosophers, so now late in my career I have wanted to be an ambassador for Chinese philosophy of technology. This requires that in China I do as much learning as teaching.

Nevertheless, one modest counsel might be to caution young Chinese students about becoming too enamored with America or the West. The American regime is much more fragile and full of contradictions than it may appear to outsiders. Additionally, the American philosophical tradition, even if it includes some elements that are worthy of emulation, is more shallow than that of China. The European philosophical tradition, with its depth and built-in on-going dynamic of criticism and recovery might serve as a better mirror in which Chinese philosophy could examine itself. This is a mirror in which it might also be useful to look when trying to think through the important modern Chinese adoption of Marx and Marxism.

ZZ: As you know, Marxism has been fundamental to the New China.

CM: And here again there may be opportunities for some learning by western scholars, adding Chinese Marxism to the broader historical development of Marxism. I think it would be fruitful for me to go back to the roots of Marxism, to consider the ideas of people such as Robert Owen, who preceded Marx, and to reconsider post-Marx European Marxists such as Rosa Luxemborg, György Lukács, Karl Korsch, and later representatives of Frankfurt School Marxism from the perspective of Chinese Marxism. You know that Andrew Feenberg has done important work to bring Frankfurt School critical social theory to bear on contemporary technology. There is also a Marxist revival going on in France in the thought of, for example, Bernard Stiegler and Thomas Piketty. Piketty's *Capital in the 21st Century* is an important study rather ignored by political philosophers of technology.

Frankfurt School Marxists were social scientists. But there is also a much under-appreciated community of British Marxists who were biological scientists. I'm thinking here especially of J.B.S. Haldane and J.D. Bernal. You know that Joseph Needham, another biologist before he re-created himself as a Sinologist, considered himself a Marxist. I sometimes imagine teaching a course in China on the British Marxists, to whom I am more attracted than continental European ones, and may have resources for come to grips with the challenges of technoscience in ways more pertinent to such problems as biodiversity loss and climate change. It would be interesting to see how students in China might respond to their work.

Once again, however, let me thank you for this stimulating exchange.

ZZ: My thanks to you as well.

References

Ames, R. T. (2011). *Confucian role ethics: A vocabulary*. Chinese University Press.

An, Y. (2005). *The idea of Cheng (sincerity/reality) in the history of Chinese philosophy*. Global Scholarly Publications.

An, Y., & Bruya, B. (Eds.). (2019). *New life for old ideas: Chinese philosophy in the contemporary world: A festscrift in honor of Donald Munro*. Chinese University Press.

Andreas, J. (2009). *Rise of the red engineers: The cultural revolution and the origins of China's new class*. Stanford University Press.

Attig, T., Bradie, M., & Rescher, N. (Eds.). (1983). *The applied turn in contemporary philosophy*. Applied Philosophy Program, Bowling Green State University.

Barmé, G. R. (2005, May). Towards a new sinology. *Chinese Studies Association of Australia Newsletter, no. 31*, 4–9. See also "What Is New Sinology?" *A New Sinology Reader*. http://chinaheritage.net/reader/what-is-new-sinology

Bray, F. (1997). *Technology and gender: Fabrics of power in late imperial China*. University of California Press.

Bray, F. (2013). *Technology, gender and history in imperial China: Great transformations reconsidered*. Routledge.

Brey, P. A. E. (2012). Anticipatory ethics for emerging technologies. *Nanoethics, 6*(1), 1–13.

Brooks, H. (1964). The scientific advisor. In R. Gilpin & C. Wright (Eds.), *Scientists and national policy making* (pp. 73–96). Columbia University Press.

Bugliarello, G., & Doner, D. (Eds.). (1979). *History and philosophy of technology*. University of Illinois Press.

Cera, A. (2020). Beyond the Empirical Turn (Elements for an Ontology of Engineering). *Információs Társadalom, 20*, 74.
Cheng, C.-Y., & Bunnin, N. (Eds.). (2002). *Contemporary Chinese philosophy*. Blackwell.
Chow, R. (1998). *Ethics after idealism: Theory, culture, ethnicity, reading*. Indiana University Press.
Cohen, P. A. (2010) *Discovering history in China: American historical writing on the recent Chinese past* (2nd rev. ed.). Columbia University Press.
Condon, R. (1959). *The Manchurian candidate*. McGraw-Hill.
Cutcliffe, S. H. (1993). Some impressions of science, technology, and society studies in China. *Technology in Society, 15*(2), 243–251.
Dasheng, G., Tsing, Z. (1989). Philosophy of Technology in China. In: Durbin, P.T. (eds) Philosophy of Technology. Philosophy and Technology, vol 6. Springer, Dordrecht. https://doi.org/10.1007/978-94-009-2303-4_8
Defoort, C., & Ames, R. T. (Eds.). (2018). *Having a word with Angus Graham: At twenty-five years into his immortality*. SUNY Press.
Farquhar, J. (2002). *Appetites: Food and sex in post-socialist China*. Duke University Press.
Farquhar, J., & Zhang, Q. (2012). *Ten thousand things: Nurturing life in contemporary Beijing*. Zone Books.
Franssen, M., Vermaas, P. E., Kroes, P., & Meijers, A. W. M. (Eds.). (2016). *Philosophy of technology after the empirical turn*. Springer.
Frodeman, R. (2006). The policy turn in environmental ethics. *Environmental Ethics, 28*(1), 3–20.
Frodeman, R., Mitcham, C., & Pielke, R., Jr. (2003). Humanities for policy—And a policy for the humanities. *Issues in Science and Technology, 20*(1), 29–32.
Frodeman, R., Holbrook, J. B., & Mitcham, C. (2012). 同行评议、研究诚信与科学治理:实践、理论与当代议题 *Tongxing pingyi, yanjiu chengxin yu kexue zhili: Shijian, lilun yu dangdai yiti/Peer review, scientific integrity, and the governance of science: Practice, theory, and current discussion [bilingual publication]*. Renmin Press.
Goodman, P. (1960). *Growing up absurd: Problems of youth in the organized system*. Vintage.
Graham, A. C. (1978). *Later Mohist logic, ethics and science*. Chinese University Press.
Graham, A. C. (1985). *Reason and spontaneity*. Cruzon Press.
Graham, A. C. (1989). *Disputers of the Tao: Philosophical argument in ancient China*. Open Court.
Guo, Q. (2018). *Studies on contemporary Chinese philosophy (1949–2009)*. Brill.
Hall, D. L., & Ames, R. T. (1987). *Thinking through Confucius*. SUNY Press.
Hall, D. L., & Ames, R. T. (1995). *Anticipating China: Thinking through the narratives of Chinese and Western culture*. SUNY Press.
Hall, D. L., & Ames, R. T. (1998). *Thinking from the Han: Self, truth, and transcendence in Chinese and Western culture*. SUNY Press.
Illich, I. (1972). *Tools for conviviality*. Harper and Row.
Illich, I., & Sanders, B. (1989). *ABC: The alphabetization of the popular mind*. Random House.
Kenner, H. (1971). *The Pound era*. University of California Press.
Lasswell, H. (1951). The policy orientation. In D. Lerner & H. Lasswell (Eds.), *The policy sciences: Recent developments in scope and method* (pp. 3–15). Stanford University Press.
Li, Z. (1988). *The path of beauty: A study of Chinese aesthetics* (L. Gong, Trans.). Oxford University Press.
Li, Z. (2010). *The Chinese aesthetic tradition* (M. B. Samei, Trans.). University of Hawai‘i Press.
Liu, D. (Ed.) (2017) 审度:马克思科学技术 观与当代科学技术论研究 *Shenduo: Makesi kexue jishu guan yu dangdai kexue jishu lun yanjiu* [Reconsiderations: A study on Marx's view and contemporary thoughts of science and technology]. Renmin University of China Press.
Liu, D., et al. (2019). 师夷长技*Shi yi zhang ji* [To learn the superior skills of the barbarians]. Renmin University of China Press.
Mitcham, C. (Ed.) (1984) Technology and philosophy. Guest edited, theme issue. *Technology in Society*, *6*(3), 187–262.
Mitcham, C. (2006). Philosophy of technology. In D. M. Borchert (Ed.), *Encyclopedia of philosophy* (Vol. 7, 2nd ed., pp. 543–551). Macmillan Reference.

Mitcham, C. (2009). A historico-ethical perspective on engineering education: From use and convenience to policy engagement. *Engineering Studies, 1*(1).

Mitcham, C. (2013). 工程与哲学:历史的、哲学的和批判的视角*Gongcheng yu zhexue: Lishide, zhexuede, he pipande shijiao* [Engineering and philosophy: Historical philosophical and critical perspectives] (Q. Wang, et al., Trans.) Renmin Press.

Mitcham, C. (Ed.) (2020) Paul Durbin and the legacy of the society for philosophy and technology. *Techné: Research in Philosophy and Technology, Special Issue, 24*(4), 1–57.

Mitcham, C., & Fisher, E. (2012). Ethics and policy. In R. Chadwick (Ed.), *Encyclopedia of applied ethics* (Vol. 2, 2nd ed., pp. 165–172). Academic Press.

Mitcham, C., & Grote, J. (1978). Philosophy of technology. In W. T. Reich (Ed.), *Encyclopedia of bioethics* (pp. 1638–1643). Free Press. Chinese adaptation: "Jishu zhexue," *Wen zhai*, vol. 8 (1989), pp. 20–23. New version in *Encyclopedia of bioethics*, 2nd ed. (1995), pp. 2477–2484 (without co-author). Revised as "Philosophy of medical technology," *Encyclopedia of bioethics*, 3rd ed. (2004), pp. 2503–2511.

Mitcham, C., & Mackey, R. (Eds.). (1972). *Philosophy and technology: Readings in the philosophical problems of technology*. Free Press.

Mitcham, C., & Mackey, R. (1973). *Bibliography of the philosophy of technology*. University of Chicago Press.

Mitcham, C., Li, B., Newberry, B., & Zhang, B. (Eds.). (2018). *Philosophy of engineering, East and West*. Springer.

Munro, D. J. (1969). *The concept of man in early China*. Center for Chinese Studies, University of Michigan.

Munro, D. J. (1977). *The concept of man in contemporary China*. Center for Chinese Studies.

Pielke, R. A., Jr. (2007). *The honest broker: Making sense of science in policy and politics*. Cambridge University Press.

Poerksen, U. (2004). *Plastic words: The tyranny of modular language* (J. Mason & D. Cayley, Trans.). Penn State University Press.

Pound, E. (1915). *Cathay*. Elkin Mathews.

Qian, Z. (Ed.). (2003). *Ezra Pound and China*. University of Michigan Press.

Rosemont, H., Jr. (Ed.). (1991). *Chinese texts and philosophical contexts: Essays dedicated to Angus C. Graham*. Open Court.

Russell, B. (1945). *A history of Western philosophy*. Simon & Schuster.

Said, E. (1978). *Orientalism*. Routledge & Kegan Paul.

Toulmin, S. (1982). How medicine saved the life of ethics. *Perspectives in Biology and Medicine, 25*, 736–750.

Verbeek, P.-P. (2005). *What things do: Philosophical reflections on technology, agency, and design*. Pennsylvania State University Press.

Wang, D. D.-w. (2017). Worlding literary China. In D. D.-w. Wang (Ed.), *A new literary history of modern China* (pp. 1–28). Harvard University Press.

Williams, R. (2002). *Retooling: A historian confronts technological change*. MIT Press.

Afterword: Continuing to Think What Might Have Gone Wrong with the West

When delivering the 2021 talk on "Re-Imagining Chinese Technological Imaginaries" that anchors this collection I never imagine it as the stimulus for such an expansive set of papers so ably edited by Levi Checketts. The original book idea came from Pieter Vermaas, but it was Checketts who picked up the idea and ran with it. To both I am sincerely indebted. To continue the dialogue thus initiated, let me venture a few further acknowledging notes.

The first must be an apology for shortcomings in the original talk. As some contributors rightly point out, its argument is simplistic, born of an enthusiasm that lacked research depth. It represented the early stage in a new research trajectory. At the talk itself, Bernadette Bensaude-Vincent, a leader in French philosophy of technology, gently pointed out the relevance (and my ignorance) of much more serious work by François Jullien. I spent the next year reading Jullien to produce a review essay. As I discovered and explained, Jullien was originally a Hellenist, but early in his career began to wonder whether the European philosophical tradition might be making assumptions about which, especially in a globalizing world, we needed to become more deeply aware. Beginning in the late 1970s, he took a "detour" (*détour*) into classical Chinese philosophy, from which, beginning in the 2000s, he "returned" to advance new understandings of the Greek and European tradition. Why China? Here is one of Jullien's own explanations from what I think is one of his most general contributions to comparative philosophy:

> Chinese and European thought have developed independently of one another for such a long time that their two worlds did not communicate (except in a very indirect way, in relation to material needs, through the Silk Road). As a matter of fact, they were completely unaware of each other until the sixteenth century (when the first evangelical missions disembarked in China, though they would have little influence) and for the most part until the nineteenth century (with the imposed opening of Chinese ports and the beginning of forced Westernization). On the one hand, the two languages, those of the European and the Chinese, are alien to one another, something which was very different from the case of India which was able to communicate with Europe through the Indo-European language (Greek

L. Checketts (ed.), *What Went Wrong with the West: Carl Mitcham's Critique of Western Philosophy of Technology*, Philosophy of Engineering and Technology 52, https://doi.org/10.1007/978-3-032-22214-5

> and Sanskrit having the same grammatical roots and categories) and, on the other hand, no historical relation of influence or contamination can be inferred between these worlds—very different again from the Islamic or Hebrew worlds which had been engaged in a continuous cultural exchange with what would become Europe: the Koran integrates Jesus as a prophet and the European Middle Ages rediscovered Aristotle through Arabic translations.
>
> At the same time, China was a world of thought just as developed, textualized, clarified and commented upon as ours in Europe. Placing China and Europe in relation would therefore be on an equal and symmetrical basis (quite differently from the dissymmetry in which anthropologists operate when they take as their terrain populations which have remained separated, but not developed writing, and are almost without history). Overall, because it is pure or not suspect, China furnished an ideal case study to use in order to test the universality of those "basic" notions we consider to be self-evident and by means of which we think. (Jullien, 2014, pp. 78–79, quoted from Mitcham, 2022, pp. 142)

For Jullien the divergence between Greek and Chinese thought and his trajectory of detour-and-return reveals an unthought in each, leaving neither simply as it was, while opening up a new "in-between." Cultures are for Jullien continuously self-transforming "fecundities," self-transforming in part through engagements with each other. My own engagement with Chinese thought about technology has similarly transformed my philosophical trajectory. Now I expect the chapters collected here to challenge me further for the remainder of my career. What might they be able to mean for anyone pursuing philosophy of technology? There is no one answer, but it is nevertheless worth speculating.

Jullien's practice exemplifies the discipline of comparative philosophy, a field of research with deeper roots in China than in the West. Initial European philosophical efforts to engage China, like those of Leibniz, Voltaire, and Hegel, were based on incredibly superficial understandings. To quote one student of contemporary comparative philosophy,

> The story is quite different in Asia, where cultural traditions mingled and clashed with considerably more frequency than in the relatively provincial West. For instance, the spread of Buddhism into China from India and central Asia beginning in the first few centuries CE sparked a long tradition of philosophical reaction to its "foreign" ideas from Confucian and Daoist intellectuals—much of it hostile, some of it appreciative and appropriating, but all of it at least implicitly comparative. The story of Chinese Buddhism over the next two millennia is very much a story of the dialogue between and among foreign and indigenous traditions, as is the story of Confucianism and Daoism during the same period.... It is, perhaps, because of this long familiarity with cross-cultural dialogue and the willingness to take one's partner seriously that many of the earliest works comparing Eastern and Western philosophies that are still important came not from Westerners but from non-Westerners responding to Western ideas. (Littlejohn, 2025, p. 3)

One of the things that might arguably be thought to have "gone wrong" in the West was its failure until recently to attempt to engage China on its own terms: a failure with consequences for thinking about technology, as variously suggested by authors here. We need more comparative philosophy of technology.

Along with Jullien, another great scholar provocateur has been Yuk Hui, whose *Question Concerning Technology in China* is referenced by Dong Luo. Luo's mapping of transitions in attitudes toward technics in the early strata of Chinese history are suggestive complements not only to Hui but also to the complementary studies of Li Zehou on the emergence of a distinctive Chinese aesthetics during the same period that are referenced in the interview with Zhihui Zhang. That interview was conducted simply to share with Chinese colleagues a naive American's intellectual trajectory and to that end was originally published in Chinese; Zhang nevertheless thought the English original belonged here.

The problematic concept of "technology" is opened further by Alfred Nordmann and Sajay Samuel. Nordmann's ambitious pursuit of a "comprehensive philosophy of technology" ventures a critical analysis, in dialogue with the under-appreciated Gernot Boehme, of the concept of "technologization" across many of its German linguistic permutations, followed with a comparative assessment of the technical visions of Francis Bacon (1561–1826) and his Chinese contemporary Song Yingxing (1587–1666). It creates a provocative contrast.

Samuel draws on the work of Ivan Illich and Giorgio Agamben to argue for the deeply Christian theological roots of Western technology, further relativizing the Western technological project. On the question of religion and technology, Agostino Cera offers a companion exploration of the religion of "dataism", drawing especially on the thinking of Günther Anders and Yuvàl Noah Harari. This is a compelling constellation. To think further about the question concerning religion and technology, I find myself trying to consider how the complex theme of the relationship between Christianity and technology might read from a Chinese perspective: well, not really from a "Chinese perspective" (which would be a claim presumptive in the extreme) but from that of an American who has been influenced by China.

Within that constellation there is, interestingly, one person, Illich, whose ideas have found resonance in China, in the cases of Tom Xiaowei Wang and Pak-Hang Wong. Both help answer my question. Wang draws on Illich (one of my great teachers) while offering an informative overview of recent developments in Chinese philosophy of technology; Wong does so when developing a creative addition to recent reinterpretations of Confucianism. Both challenge any superficial ideas about Chinese philosophy of technology and invite comparative philosophical reflections about how philosophy of technology might manifest as same or different in China and the West.

In an original approach related to that kind on inquiry, Levi Checketts draws on Jacques Lacan to develop an analysis of ways different cultural self-constituting experiences of desire and anxiety set them up for or catalyze attractions to technology. In a chiropractic on the question about what went wrong with the West, Checketts wants to ask why does anyone, East or West, "want to be like the West"? and "How can we transcend this instrumental rational approach to comparative philosophy?" These are complementary or better and more destabilizing questions than my own.

However, in one of the most destabilizing pieces in the collection the almost monograph-length piece by Christiopher Coenen and Aleksandra Kazakova reframes the whole West/East problematic by bringing Russia into the picture. Russia, with its own distinctive uptake of desire, anxiety, and technology is at once a more immediate West to China than Europe and a more immediate East to Europe than China. One begins to wonder about what went wrong with understandings of the self and other everywhere with regard to desire and anxiety in the appropriations of modern engineering and technology.

To conclude my note hypothesizing this volume as part of a work-in-progress in comparative philosophy of technology, let me quote from the opening paragraphs of David Wong's (2024) *Stanford Encyclopedia of Philosophy* entry on "Comparative Philosophy: Chinese and Western":

> Comparative philosophy originated from bringing together philosophical traditions that are perceived to have developed in relative isolation from one another and that are defined quite broadly along cultural and regional lines—Chinese versus Western, for example.... One benefit of comparative philosophy lies in the way that it forces reflection on the most deeply entrenched and otherwise unquestioned agendas and assumptions of one's own tradition.... As the field has matured, discussion has expanded beyond ... commensurability questions of comparison to the question, "What can we learn from other traditions?" and even "Is it now time to regard ourselves as members of more than one tradition or as members of hybrid traditions?" To the extent that different traditions address common or overlapping problems, and as some comparative philosophers become more solidly grounded in more than one tradition, we might expect better and more original philosophical work from those who can draw from a greater range of traditions.

References

Jullien, F. (2014). *On the universal, the uniform, the common and dialogue between cultures* (M. Richardson & K. Fijalkowski, Trans.). Polity.

Littlejohn, R. (2025). Comparative philosophy. *The Internet Encyclopedia of Philosophy*, ISSN 2161-0002. https://iep.utm.edu/comparative-philosophy

Mitcham, C. (2022). Review of three books by or on François Jullien. *Dao, 21*(February), 141–150.

Wong, D. (2024). Comparative philosophy: Chinese and Western. In E. N. Zalta & U. Nodelman (Eds.) *Stanford encyclopedia of philosophy* (Fall 2024 Edition). https://plato.stanford.edu/archives/fall2024/entries/comparphil-chiwes/

Index

L. Checketts (ed.), *What Went Wrong with the West: Carl Mitcham's Critique of Western Philosophy of Technology*, Philosophy of Engineering and Technology 52, https://doi.org/10.1007/978-3-032-22214-5

MIX
Papier aus verantwortungsvollen Quellen
Paper from responsible sources
FSC® C105338

If you have any concerns about our products,
you can contact us on
ProductSafety@springernature.com

In case Publisher is established outside the EU,
the EU authorized representative is:
Springer Nature Customer Service Center GmbH
Europaplatz 3, 69115 Heidelberg, Germany

Printed by Libri Plureos GmbH
in Hamburg, Germany